An Invitation to Model Theory

Model theory begins with an audacious idea: to consider statements about mathematical structures as mathematical objects of study in their own right. While inherently important as a branch of mathematical logic, it also enjoys connections to and applications in diverse branches of mathematics, including algebra, number theory and analysis. Despite this, traditional introductions to model theory assume a graduate-level background of the reader.

In this innovative textbook, Jonathan Kirby brings model theory to an undergraduate audience. The highlights of basic model theory are illustrated through examples from specific structures familiar from undergraduate mathematics, paying particular attention to definable sets throughout. With numerous exercises of varying difficulty, this is an accessible introduction to model theory and its place in mathematics.

JONATHAN KIRBY is a Senior Lecturer in Mathematics at the University of East Anglia. His main research is in model theory and its interactions with algebra, number theory, and analysis, with particular interest in exponential functions. He has taught model theory at the University of Oxford, the University of Illinois at Chicago, and the University of East Anglia.

An Invitation to Model Theory

JONATHAN KIRBY
University of East Anglia

CAMBRIDGE
UNIVERSITY PRESS

University Printing House, Cambridge CB2 8BS, United Kingdom

One Liberty Plaza, 20th Floor, New York, NY 10006, USA

477 Williamstown Road, Port Melbourne, VIC 3207, Australia

314–321, 3rd Floor, Plot 3, Splendor Forum, Jasola District Centre,
New Delhi – 110025, India

79 Anson Road, #06–04/06, Singapore 079906

Cambridge University Press is part of the University of Cambridge.

It furthers the University's mission by disseminating knowledge in the pursuit of education, learning, and research at the highest international levels of excellence.

www.cambridge.org
Information on this title: www.cambridge.org/9781107163881
DOI: 10.1017/9781316683002

First published 2019

Printed and bound in Great Britain by Clays Ltd, Elcograf S.p.A.

A catalogue record for this publication is available from the British Library.

Library of Congress Cataloging-in-Publication Data
Names: Kirby, Jonathan, 1979– author.
Title: An invitation to model theory / Jonathan Kirby, University of East Anglia.
Description: Cambridge, United Kingdom ; New York, NY : Cambridge University Press, 2019. | Includes bibliographical references and index.
Identifiers: LCCN 2018052996 | ISBN 9781107163881 (hardback ; alk. paper) | ISBN 1107163889 (hardback ; alk. paper) | ISBN 9781316615553 (pbk. ; alk. paper) | ISBN 1316615553 (pbk. ; alk. paper)
Subjects: LCSH: Model theory.
Classification: LCC QA9.7 .K57 2019 | DDC 511.3/4–dc23
LC record available at https://lccn.loc.gov/2018052996

ISBN 978-1-107-16388-1 Hardback
ISBN 978-1-316-61555-3 Paperback

To Pirita, Lumia, Tapio, and Sakari

Contents

Preface

This book is designed as an undergraduate or master's-level course in model theory. It has grown out of courses taught for many years at the University of Oxford and courses taught by me at UEA. The choice of material and presentation are based on pedagogical considerations, and I have tried to resist the temptation to be encyclopedic.

In this book, the main programme of model theory is to take a familiar mathematical structure and get an understanding of it in the following way. First, find an axiomatisation of its complete theory. Second, if possible, classify all the other models of the theory. Third, describe all the definable sets. As a result, model theory is presented as a set of tools for understanding structures, and the way the tools are applied to specific structures is as important as the tools themselves. This gives motivation to the subject and connects it to familiar material. Some readers may be more interested in the theory than in the applications, but my view is that even those who eventually wish to work in abstract model theory will get a better understanding of the basics by seeing them applied to examples.

Historically, model theory grew as a branch of mathematical logic, and the focus was mostly on logical issues, such as decidability. As model theory has found more applications and connections to other branches of mathematics, the study of definable sets has become more central.

I have tried to keep the book as self-contained as possible. Model theory requires a level of mathematical sophistication in terms of abstract, rigorous thinking, proofs, and algebraic thinking which students will normally have developed through previous courses in algebra, logic, or geometry, but there are few specific prerequisites. No topology is used, and almost no set theory is used. A brief chapter explains the basic cardinal arithmetic methods needed for some of the proofs, but with one or two exceptions, everything can be assumed to be countable. A familiarity with the use of basic algebraic ideas, such as bijections and homomorphisms or embeddings of groups, or rings, or vector spaces, is essential, but when algebraic examples such as rings and vector spaces are introduced, all the necessary definitions and facts are explained.

For the most part I have not given historical references for the material. There are good historical remarks at the end of each chapter in the books of Hodges [Hod93, Hod97] and Marker [Mar02], and I refer the reader to those.

Overview

The book is organised into six parts. Part I covers structures, languages, and automorphisms, introduces definable sets, and proves the essential preservation theorems.

Part II introduces theories and the programme of finding axiomatisations for structures. Context is given with axiomatisations of the complex and real fields and the natural numbers with addition and multiplication (Peano arithmetic). The compactness theorem is then introduced, and examples of its use are given. Part II concludes with the Henkin proof of the compactness theorem.

In Part III we show that even a complete theory can have many different models with the Löwenheim–Skolem theorems. The notion of categoricity is introduced via the example of vector spaces and is used to prove the completeness of an axiomatisation. Further applications are given to dense linear orders, where the back-and-forth method is introduced, and to the natural numbers with the successor function.

The idea of quantifier elimination is introduced in Part IV, and used to characterise the definable sets in dense linear orders and vector spaces. Boolean algebras are introduced via an investigation into the theory of power sets, partially ordered by the subset relation, and the definable subsets of a structure in any number n of variables is shown to be a Boolean algebra, called the Lindenbaum algebra of the theory. These Lindenbaum algebras are then worked out in the examples of vector spaces and for the real ordered field.

Parts V and VI go in different directions and do not depend on each other. Part V develops the notion of types, which is at the heart of modern model theory. The first goal is the Ryll–Nardzewski theorem, which characterises countably categorical theories as those for which there are only finitely many definable sets in any given number of variables. There is then a brief discussion of saturated models, leading to areas for further reading.

Part VI takes the model-theoretic techniques developed in the first four parts and applies it to the theory of algebraically closed fields. Two chapters explain the necessary algebraic background, and a third proves categoricity, completeness of the theory, and quantifier elimination. The next chapter explains how the definable sets correspond to algebraic varieties and constructible sets, and a final chapter gives a model theoretic proof of Hilbert's Nullstellensatz, which gives more information about the definable sets.

Suggestions for Using This Book as a Textbook

This book can be used for a course in many different ways. Figure 0.1, illustrating the dependencies between chapters, can be used as a guide to planning a course according to the lecturer's preferences.

Each chapter is intended to be of such a length that it can be taught in one hour (perhaps covering only the essentials) or in two hours (sometimes using material from the exercises). Everything depends on the first three chapters, so they should be covered first, unless the students have the prerequisite knowledge from predicate logic. The emphasis is on semantic ideas, including automorphisms, which is somewhat different from the usual emphasis in a first logic course, and I have successfully taught this material to a class of students of whom some had seen logic before and some not. For those who have not seen this material before, Chapter 3 in particular goes rather quickly. Chapters 4 and 5 carry on the study of formulas to give students more opportunity to consolidate their understanding.

Each chapter has several exercises at the end, which range from very easy consolidations of definitions to more substantial projects. Some exercises need more background than the other material in the chapter, and some exercises develop extension material.

A short model theory section of a mathematical logic course could have the compactness theorem and some applications as its goal and consist of Chapters 1, 2, 3, and 6, Section 8.1, and Chapter 9. This could be filled out with any of Chapters 4, 5, and 7, the rest of Chapter 8 or Chapter 11, or with applications from Chapters 14, 15, 16, or 19. The back-and-forth method is another highlight that could be reached quickly, by covering only the essential sections of Chapters 1, 3, 6, and 17.

The heart of this book is the study of definable sets. A course centred on those would consist of Parts I to IV, possibly omitting Chapters 5, 9, 11, and 16, and possibly adding Part VI.

A course aiming to cover the basics of types and the Ryll–Nardzewski theorem, without the emphasis on definable sets, could consist of Chapters 1, 2, 3, 6, 8, 10–15, and 23–25.

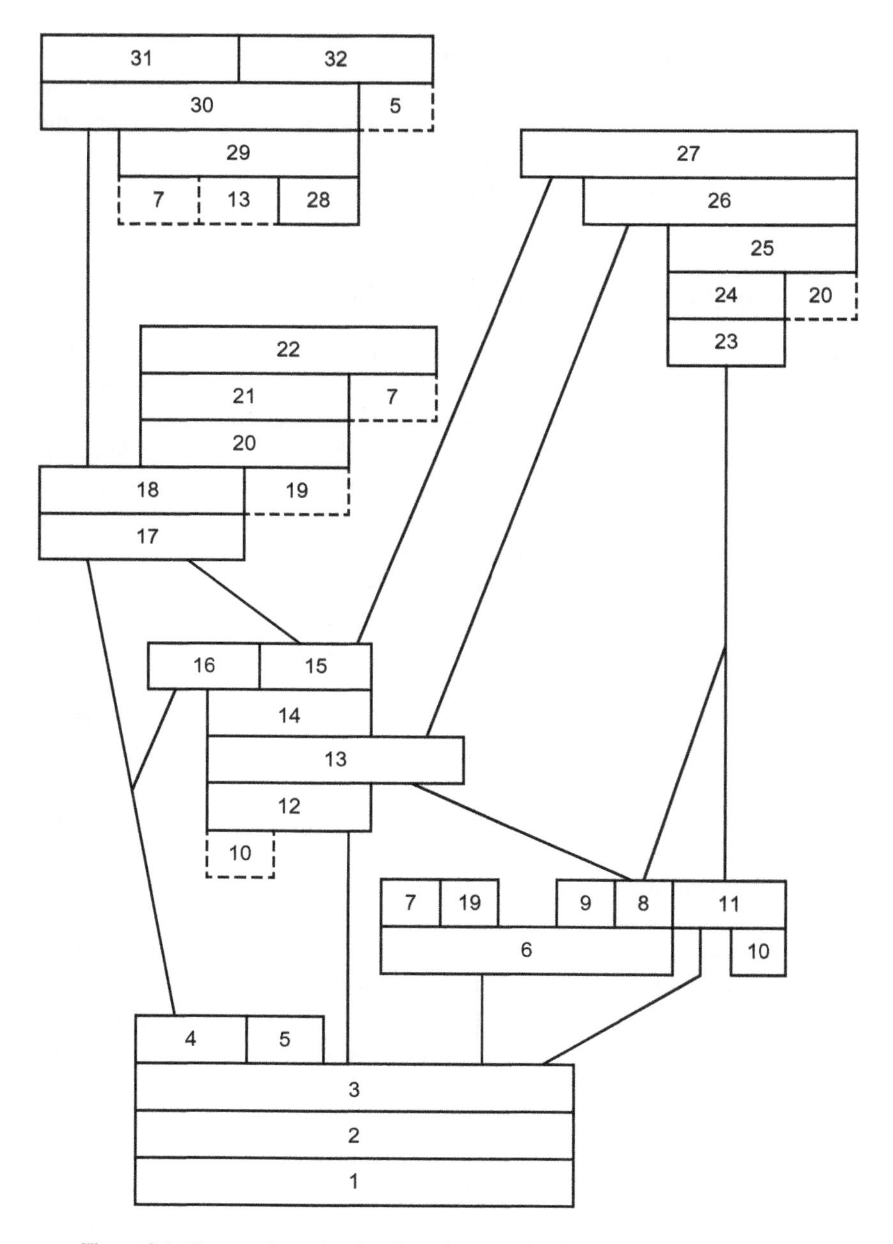

Figure 0.1 Chapter dependencies: In order to make the diagram planar, Chapters 5, 7, 10, 13, 19, and 20 have been drawn in two (or even three) places. As the diagram shows, Parts IV, V, and VI are largely independent of each other, with much of Part V also independent of Part III. While Chapter 25 uses the concept of Lindenbaum algebras from Chapter 20, that material does not rely on the rest of Parts III or IV.

For research students, Part VI, on algebraically closed fields, gives a pattern for carrying out the programme of understanding definable sets in a theory which is similar to the pattern which works in many other theories of fields, such as real-closed fields, differentially closed fields, separably closed fields, algebraically closed valued fields, exponentially closed fields, algebraically closed fields with an automorphism, and so on.

Further Reading

There are several more advanced textbooks in model theory, including those of Marker [Mar02], Hodges [Hod93, Hod97], Poizat [Poi00], Tent and Ziegler [TZ12], Sacks [Sac10], and Chang and Keisler [CK90]. Rothmaler [Rot00] is more at the level of this book, but there is more emphasis on algebra, especially groups. Väänänen [Vää11] introduces model theory via back-and-forth games, which complements the approach in this book. Bridge [Bri77] was based on the incarnation of the Oxford model theory course from the early 1970s, and the emphasis is much more on logic. Other suggestions for further reading are given through the book, particularly at the end of Parts IV, V, and VI.

Acknowledgements

Thanks to Boris Zilber for allowing me to take his lecture notes as a starting point both for my course and for this book. I learned model theory initially from the books of Wilfrid Hodges and David Marker, and from Boris. My writing style was much influenced by the late Harold Simmons.

I would like to thank Cambridge University Press for their help in producing the book and, in particular, Silvia Barbina for initially suggesting that I write it.

Many people have made helpful comments on drafts or have otherwise helped me to write the book. I would like to thank Lou van den Dries, David Evans, Åsa Hirvonen, Wilfrid Hodges, Ehud Hrushovski, Tapani Hyttinen, Gareth Jones, Asaf Karagila, David Marker, Alice Medvedev, Charles Steinhorn, Alex Wilkie, and Boris Zilber for sharing their expertise. Thanks to Francesco Parente for reading two drafts and providing many considered comments.

Thanks also to the students who attended my courses or read drafts of the book and provided essential feedback. They include Abeer Albalahi, Michael Arnold, Emma Barnes, Matt Gladders, Grant Martin, Oliver Matheau-Raven, Audie Warren, and Tim Zander.

Part I

Languages and Structures

In this first part of the book, we introduce the basic methods from predicate logic which underpin the rest of the book. Mathematical objects are formalised as L-structures, and statements about these structures are formalised in the notions of L-terms and L-formulas. We introduce the central notion of a definable set, a subset of an L-structure which is defined by an L-formula. A key notion in mathematical logic is that of a recursive definition, and both the notions of L-terms and L-formulas give examples of that. These recursive definitions give rise to methods of proof by induction. Throughout this part, these proofs by induction are used prove preservation theorems, most importantly the preservation of definable sets by automorphisms of a structure. From this result we get a method to show that certain subsets of a structure are *not* definable, an essential tool which we will later exploit in the programme of characterising the definable sets of a structure.

1
Structures

In different mathematical contexts, familiar mathematical objects may actually have different meanings. For example a reference to the integers $\mathbb{Z}$ in group theory is likely to mean the infinite cyclic group, whereas in number theory it is likely to mean $\mathbb{Z}$ as a ring. In model theory, when we specify a structure, we have to be precise about such things. The integers as an additive group will be written as $\mathbb{Z}_{\text{adgp}} = \langle \mathbb{Z}; +, -, 0 \rangle$, and $\mathbb{Z}$ as a ring will be written as $\mathbb{Z}_{\text{ring}} = \langle \mathbb{Z}; +, \cdot, -, 0, 1 \rangle$. The integers are also used to index the years in a calendar, and an appropriate structure for that purpose is $\mathbb{Z}_{<} = \langle \mathbb{Z}; < \rangle$, because it is not very meaningful to add or multiply calendar years, but the order is important. In general, we capture these ideas with the notions of a language, L, and an L-structure.

1.1 L-structures

Definition 1.1 A *language* L is specified by the following (sometimes called the *vocabulary* or *signature* of L):

(i) a set of *relation symbols*,
(ii) a set of *function symbols*,
(iii) a set of *constant symbols*, and
(iv) for each relation and function symbol, a positive natural number called its *arity*.

Definition 1.2 An *L-structure* $\mathcal{A}$ consists of a set A called the *domain* of the structure, together with *interpretations* of the symbols from L:

(i) for each relation symbol R of L, of arity n, a subset $R^{\mathcal{A}}$ of A^n,
(ii) for each function symbol f of L, of arity n, a function $f^{\mathcal{A}} : A^n \to A$, and
(iii) for each constant symbol c of L an element $c^{\mathcal{A}} \in A$.

Example 1.3 The language of rings $L_{\text{ring}} = \langle +, \cdot, -, 0, 1 \rangle$, where $+$ and $\cdot$ are function symbols of arity 2 (we say they are *binary* function symbols), $-$ is a function symbol of arity 1 (a *unary* function symbol for negation rather than the binary function of subtraction), and 0 and 1 are constant symbols. $\mathbb{Z}_{\text{ring}}$ is the L_{ring}-structure with domain the set $\mathbb{Z}$ of integers. We interpret the symbols $+$, $\cdot$, and $-$ as the usual functions of addition, multiplication, and negation of integers and the constant symbols 0 and 1 are interpreted as the usual zero and one. We can write this as $\mathbb{Z}_{\text{ring}} = \langle \mathbb{Z}; +^{\mathbb{Z}}, \cdot^{\mathbb{Z}}, -^{\mathbb{Z}}, 0^{\mathbb{Z}}, 1^{\mathbb{Z}} \rangle$ if for example we want to distinguish the symbol $+$ from its interpretation $+^{\mathbb{Z}}$ as a function from $\mathbb{Z}^2$ to $\mathbb{Z}$. This distinction can be important in mathematical logic. For example we also have the L_{ring}-structure $\mathbb{R}_{\text{ring}}$ with domain the set of real numbers $\mathbb{R}$, and the function $+^{\mathbb{R}} : \mathbb{R}^2 \to \mathbb{R}$ is not the same function as $+^{\mathbb{Z}} : \mathbb{Z}^2 \to \mathbb{Z}$. However, usually no ambiguity arises, and we just write $+$ for the symbol and for its interpretations in different structures.

Example 1.4 We write $L_< = \langle < \rangle$, where $<$ is a binary relation symbol. Then $\mathbb{Z}_<$ is the $L_<$-structure with domain the set $\mathbb{Z}$ of integers, and the symbol $<$ is interpreted as the set of pairs $\left\{(a, b) \in \mathbb{Z}^2 \mid a < b\right\}$. As above, we could write this set as $<^{\mathbb{Z}}$, but usually we will not.

Examples 1.5 Many other languages can be built as variations on these two languages:

(i) $L_{\text{gp}} = \langle \cdot, (-)^{-1}, 1 \rangle$ is the language of groups. Again, $\cdot$ is a binary function symbol, $(-)^{-1}$ is a unary function symbol representing the multiplicative inverse function $x \mapsto x^{-1}$, and 1 is a constant symbol.

(ii) $L_{\text{adgp}} = \langle +, -, 0 \rangle$ is the language of groups written additively. It is a *sub-language* of L_{ring}, because every symbol in L_{adgp} is also in L_{ring} (with the same arity). We also say that L_{ring} is an *expansion* of the language L_{adgp}.

(iii) $L_{\text{o-ring}} = \langle +, \cdot, -, 0, 1, < \rangle$ is the language of ordered rings, consisting of $L_{\text{ring}} \cup L_<$.

(iv) A common language in which to consider the natural numbers is the language of semirings, $L_{\text{s-ring}} = \langle +, \cdot, 0, 1 \rangle$. We will always use the convention that 0 is a natural number.

(v) The language of monoids is $L_{\text{mon}} = \langle \cdot, 1 \rangle$ and the language of additive monoids is $L_{\text{admon}} = \langle +, 0 \rangle$.

Most of the structures we will consider as examples have a domain with a commonly used notation such as $\mathbb{Z}$, $\mathbb{Q}$, or $\mathbb{R}$, and the functions, relations, and constants we consider will be those in the above languages. In this case we name the structure by putting the name of the language as a subscript, for example $\mathbb{Z}_{\text{ring}}$, $\mathbb{R}_{\text{o-ring}}$, $\mathbb{Q}_{<}$, $\mathbb{N}_{\text{s-ring}}$.

We can also specify structures directly, which is useful for creating simpler examples. In this case we will often use a caligraphic letter for the name of the structure and the corresponding Roman letter for the name of its domain, so $\mathcal{A}$ would be the name of a structure with domain A. This convention of using different notation for the name of a structure and its domain is useful when the same set is the domain of different structures. However, where no confusion is likely to arise, we will sometimes follow the common mathematical practice of using the same notation for both.

Example 1.6 Take the language $L = \langle R, f, c \rangle$, where R is a ternary relation symbol, f is a unary function symbol, and c is a constant symbol. We can define an L-structure $\mathcal{A}$ with domain $A = \{0, 1, 2, 3, 4\}$, by specifying the interpretation of the symbols as $f^{\mathcal{A}}(x) = x+1 \mod 5$, $R^{\mathcal{A}} = \left\{(x, y, z) \in A^3 \,\middle|\, \text{exactly two of } x, y \text{ and } z \text{ are equal}\right\}$ and $c^{\mathcal{A}} = 3$.

Remark 1.7 The key idea from mathematical logic here is that the symbols of the vocabulary of a language L are separate from their interpretations in a structure. It follows that the symbols do not have to be interpreted by their usual meanings. For example we can make the set $\mathbb{N}$ into an L_{gp}-structure by interpreting the symbol $\cdot$ as addition and $(-)^{-1}$ as the identity function, so $x^{-1} = x$ for all $x \in \mathbb{N}$, and interpreting the constant symbol 1 as the number 2. However, this is perverse, and if we want an unusual interpretation, we will choose to use a different symbol.

Given a mathematical problem you are trying to solve, or a statement you want to understand, it is often a good exercise to work out what structure the statement might be about. For example, the fundamental theorem of arithmetic states that every positive integer can be written as a product of primes in a unique way (up to reordering). An appropriate structure would have domain the set $\mathbb{N}^+$ of positive integers and needs to have the multiplication function; 1 is a special case as the empty product, which is relevant to single out, so we can regard the fundamental theorem of arithmetic as a statement about the structure $\mathbb{N}^+_{\text{mon}} = \langle \mathbb{N}^+; \cdot, 1 \rangle$. To actually prove the theorem, we might need more than that, for example the order to do induction on and also addition.

1.2 Expansions and Reducts

When considering different structures with the same domain, there are two useful pieces of terminology.

Definition 1.8 Let L be a language and L^+ another language such that $L \subseteq L^+$, that is, every symbol of L is also a symbol of L^+. Let $\mathcal{A}^+$ be an L^+-structure with domain A. Then the *reduct* of $\mathcal{A}^+$ to L is the L-structure $\mathcal{A}$ with domain A, and every symbol of L interpreted in $\mathcal{A}$ exactly as in $\mathcal{A}^+$. We also say that $\mathcal{A}^+$ is an *expansion* of $\mathcal{A}$ to the language L^+.

For example $\mathbb{R}_{\text{adgp}} = \langle \mathbb{R}; +, -, 0 \rangle$ is a reduct of $\mathbb{R}_{\text{ring}} = \langle \mathbb{R}; +, -, \cdot, 0, 1 \rangle$, which in turn is a reduct of $\mathbb{R}_{\text{o-ring}} = \langle \mathbb{R}; +, -, \cdot, 0, 1, < \rangle$.

1.3 Embeddings and Automorphisms

Definition 1.9 Let $\mathcal{A}$ and $\mathcal{B}$ be L-structures. An *embedding* of L-structures from $\mathcal{A}$ to $\mathcal{B}$ is an injective function $A \xrightarrow{\pi} B$ such that:

(i) for all relation symbols R of L, and all $a_1, \ldots, a_n \in A$,

$$(a_1, \ldots, a_n) \in R^{\mathcal{A}} \text{ iff } (\pi(a_1), \ldots, \pi(a_n)) \in R^{\mathcal{B}},$$

(ii) for all function symbols f of L, and all $a_1, \ldots, a_n \in A$,

$$\pi(f^{\mathcal{A}}(a_1, \ldots, a_n)) = f^{\mathcal{B}}(\pi(a_1), \ldots, \pi(a_n)), \quad \text{and}$$

(iii) for all constant symbols c of L, $\pi(c^{\mathcal{A}}) = c^{\mathcal{B}}$.

For any L-structure $\mathcal{A}$, the identity function 1_A on A is a embedding of $\mathcal{A}$ into itself. An embedding $\mathcal{A} \xrightarrow{\pi} \mathcal{B}$ is an *isomorphism* iff there is an embedding $\mathcal{B} \xrightarrow{\sigma} \mathcal{A}$ such that the composite $\pi \circ \sigma$ is the identity on $\mathcal{B}$ and the composite $\sigma \circ \pi$ is the identity on $\mathcal{A}$. We write π^{-1} for σ, as usual, and call it the inverse of π. An isomorphism from $\mathcal{A}$ to itself is called an *automorphism* of $\mathcal{A}$.

Examples 1.10 We have the obvious inclusion functions $\mathbb{Z} \hookrightarrow \mathbb{Q} \hookrightarrow \mathbb{R} \hookrightarrow \mathbb{C}$, which take a number to itself. These inclusion functions give us embeddings of L_{ring}-structures

$$\mathbb{Z}_{\text{ring}} \hookrightarrow \mathbb{Q}_{\text{ring}} \hookrightarrow \mathbb{R}_{\text{ring}} \hookrightarrow \mathbb{C}_{\text{ring}}.$$

Definition 1.11 If we have an embedding $\mathcal{A} \longrightarrow \mathcal{B}$ where the function is an inclusion of the domain of $\mathcal{A}$ as a subset of the domain of $\mathcal{B}$ then we say that $\mathcal{A}$ is a *substructure* of $\mathcal{B}$, and that $\mathcal{B}$ is an *extension* of $\mathcal{A}$.

Example 1.12 The only automorphisms of $\mathbb{Z}_{\text{adgp}}$ are the identity and the map $x \mapsto -x$. To see this, note that if $\pi : \mathbb{Z} \to \mathbb{Z}$ is an L_{adgp}-embedding, then $\pi(0) = 0$, and if $\pi(1) = n$, then for any $m \in \mathbb{N}^+$, we have

$$\pi(m) = \pi(\underbrace{1 + \cdots + 1}_{m}) = \underbrace{\pi(1) + \cdots + \pi(1)}_{m} = mn.$$

Then also $\pi(-n) = -\pi(n) = -mn$. To have an inverse, π must be surjective, which implies $n = \pm 1$.

Exercises

1.1 Let $L = \langle f, c \rangle$ be a language with one unary function symbol and one constant symbol. Describe all the possible L-structures on the domain $\{1, 2\}$. How many of them are there up to isomorphism (which means counting isomorphic structures as the same)?

1.2 For each of the following statements, give a structure which they say something about.

(a) There is no largest integer.
(b) Every integer has an additive inverse.
(c) Square integers are always non-negative.
(d) Every complex number is the sum of a real number and i times a real number.
(e) The exponential function is strictly increasing on the real numbers.
(f) A real quadratic equation $ax^2 + bx + c = 0$ has real roots if and only if $b^2 - 4ac \geqslant 0$.
(g) Euler's identity $e^{i\pi} + 1 = 0$.

1.3 For each language L in Examples 1.5, say which of the sets from $\mathbb{N}$, $\mathbb{Z}$, and $\mathbb{R}$ is the domain of an L-structure with the symbols interpreted with their usual meanings. For example, there is no structure $\mathbb{N}_{gp}$ because $\mathbb{N}$ is not closed under multiplicative inverses. Which of your structures are expansions or reducts of each other? Which are extensions or substructures?

1.4 What are all the automorphisms of $\mathbb{Z}_<$?

1.5 Show that embeddings of $\mathbb{N}_<$ into $\mathbb{R}_<$ correspond to strictly increasing sequences of real numbers.

1.6 Show that an embedding of L-structures is an isomorphism if and only if it is surjective.

1.7 Explain what all the embeddings of $\mathbb{Z}_{adgp}$ into $\mathbb{R}_{adgp}$ are.

1.8 Find an automorphism π of the structure $\langle \mathbb{R}; < \rangle$ such that $\pi(0) = 1$ and $\pi(1) = 5$. Is your π also an automorphism of the structure $\langle \mathbb{R}; + \rangle$?

1.9 Find all the automorphisms of $\mathbb{Q}_{\text{adgp}}$.

1.10 Suppose that $\mathcal{A}^+$ is an expansion of $\mathcal{A}$ and π is an automorphism of $\mathcal{A}^+$. Show that π is also an automorphism of $\mathcal{A}$. Give an example to show that an automorphism of $\mathcal{A}$ may not be an automorphism of $\mathcal{A}^+$.

1.11 Let $\mathcal{A}$ be an L-structure. Show that the automorphisms of $\mathcal{A}$ form a group with the group operation being composition. We write this group as $\text{Aut}(\mathcal{A})$.

1.12 In model theory, a *homomorphism* of L-structures $\mathcal{A} \xrightarrow{\pi} \mathcal{B}$ is a function $A \xrightarrow{\pi} B$ such that

(a) for all relation symbols R of L, and all $a_1, \ldots, a_n \in A$, if $(a_1, \ldots, a_n) \in R^{\mathcal{A}}$ then $(\pi(a_1), \ldots, \pi(a_n)) \in R^{\mathcal{B}}$,
(b) for all function symbols f of L, and all $a_1, \ldots, a_n \in A$, $\pi(f^{\mathcal{A}}(a_1, \ldots, a_n)) = f^{\mathcal{B}}(\pi(a_1), \ldots, \pi(a_n))$, and
(c) for all constant symbols c of L, $\pi(c^{\mathcal{A}}) = c^{\mathcal{B}}$.

Check that if $L = L_{\text{gp}}$ or $L = L_{\text{ring}}$, and $\mathcal{A}, \mathcal{B}$ are groups or rings, then L-homomorphisms are the same as group or ring homomorphisms. How do L-homomorphisms compare with, and how do they differ from, the notion of embedding of L-structures?

2
Terms

A formal language is based on a collection of symbols, which consists of variables such as x, y, or z, logical symbols such as $=$, $\wedge$, $\neg$, and $\exists$, brackets (,), [, and], and then the vocabulary which is specific to the language: the relation, function, and constant symbols. Apart from the collection of symbols, a language also consists of various strings (lists) of the symbols, which are called terms and formulas. These symbols, terms, and formulas are then interpreted in structures. In the previous chapter we interpreted the symbols from the vocabulary, and in the next chapter we will introduce the formulas. Here we introduce and interpret the terms.

2.1 The Recursive Construction of Terms

Terms are certain strings of symbols that refer to elements of the structure or to functions on the structure. For example, in the language L_{ring}, terms include

$$0, \quad (1+1), \quad -((1+1)+1), \quad (x \cdot (1+1)), \quad (x+y), \text{ and } ((x \cdot x)+1).$$

They can be interpreted in $\mathbb{Z}_{\text{ring}}$ as the numbers 0, 2, and -3 and as the functions $x \mapsto 2x$, $(x, y) \mapsto x + y$, and $x \mapsto x^2 + 1$, respectively.

Definition 2.1 (Terms) The set of *terms* of the language L is defined recursively as follows.

(i) Every variable is a term.
(ii) Every constant symbol of L is a term.
(iii) If f is a function symbol of L of arity n, and $t_1, \ldots, t_n$ are terms of L, then $f(t_1, \ldots, t_n)$ is a term.
(iv) Only something built from the above three clauses in finitely many steps is a term.

A *closed term* is a term which does not contain any variables.

In the examples we followed the usual practice of writing $(1 + 1)$ rather than $+(1, 1)$ and so on. This is called *infix notation* and works only for binary functions. When the function is written before its arguments, it is called *prefix notation*.

As in the examples above, terms of L can be interpreted as elements of an L-structure or functions on it.

Definition 2.2 (Interpretation of terms) Let $\mathcal{A}$ be an L-structure, t a term of L, and $\bar{x} = (x_1, \dots, x_r)$ a list of variables including all those which appear in t. The recursive definition of terms is used to give a recursive definition of a function $t^{\mathcal{A}} : A^r \to A$ as follows.

(i) If $t = c$, a constant symbol, then $t^{\mathcal{A}}$ is the element $c^{\mathcal{A}}$ of A, or as a function it is the constant function which always takes the value $c^{\mathcal{A}}$.
(ii) If $t = x_i$, the ith variable in the list $\bar{x}$, then $t^{\mathcal{A}}$ is the ith coordinate function given by $t^{\mathcal{A}}(x_1, \dots, x_r) = x_i$.
(iii) If $t = f(t_1, \dots, t_n)$, then $t^{\mathcal{A}}$ is given by composition

$$t^{\mathcal{A}}(\bar{x}) = f^{\mathcal{A}}(t_1^{\mathcal{A}}(\bar{x}), \dots, t_n^{\mathcal{A}}(\bar{x})).$$

Remarks 2.3 (i) If t is a closed term, then in Definition 2.2 we can take $r = 0$. In this case it is best to think of the interpretation $t^{\mathcal{A}}$ as an element of A. We can alternatively think of it as a function $A^0 \to A$. Since A^0 is a one-point set, this amounts to the same thing as an element of A.
(ii) Note that terms depend only on the language, but their interpretations depend on the structure. For example, the term $((x + 1) + x)$ is interpreted in the structure $\mathbb{R}_{\text{ring}}$ as the function

$$\begin{aligned} \mathbb{R} &\longrightarrow \mathbb{R} \\ x &\longmapsto 2x + 1, \end{aligned}$$

but in $\mathbb{Z}_{\text{ring}}$ the same term is interpreted as a function $\mathbb{Z} \to \mathbb{Z}$.
(iii) We can see that several terms can be interpreted as the same function. For example, the terms $((-x + (x + x)) + (1 + 1))$ and $((1 + x) + 1)$ are both interpreted in $\mathbb{Z}_{\text{ring}}$ as the function

$$\begin{aligned} \mathbb{Z} &\longrightarrow \mathbb{Z} \\ x &\longmapsto x + 2. \end{aligned}$$

(iv) Where no ambiguity arises, we often do not write all the brackets which should be there according to the recursive definition.

(v) In model theory we almost always deal with terms with a given list of variables. So we will write that $t(\bar{x})$ is a term, when we really mean that t is a term and $\bar{x}$ is a list of variables which includes all the variables which appear in t.

2.2 Embeddings and Terms

By definition, an embedding of L-structures preserves the interpretations of the constant and function symbols from L. We can use the recursive definition of terms to give an inductive proof that embeddings *preserve* the interpretation of all terms. When $\bar{a} = (a_1, \ldots, a_r)$ is a tuple of elements from A, we write $\pi(\bar{a})$ as an abbreviation for the tuple $(\pi(a_1), \ldots, \pi(a_r))$. Otherwise the notation becomes too unwieldy.

Proposition 2.4 *Suppose $\mathcal{A} \xrightarrow{\pi} \mathcal{B}$ is an embedding of L-structures, t is a term of L, $\bar{x} = (x_1, \ldots, x_r)$ is a list of variables including all those which appear in t, and $\bar{a} \in A^r$. Then*

$$\pi(t^{\mathcal{A}}(\bar{a})) = t^{\mathcal{B}}(\pi(\bar{a})).$$

Proof We proceed by induction on the construction of terms.

If t is a constant symbol c, then

$$\begin{aligned} \pi(t^{\mathcal{A}}(\bar{a})) &= \pi(c^{\mathcal{A}}) && \text{by Definition 2.2,} \\ &= c^{\mathcal{B}} && \text{by the definition of an embedding,} \\ &= t^{\mathcal{B}}(\pi(\bar{a})) && \text{by Definition 2.2 again.} \end{aligned}$$

If $t = x_i$, then $\pi(t^{\mathcal{A}}(\bar{a})) = \pi(a_i) = t^{\mathcal{B}}(\pi(\bar{a}))$, using Definition 2.2 twice.
If $t = f(t_1, \ldots, t_s)$, then

$$\begin{aligned} \pi(t^{\mathcal{A}}(\bar{a})) &= \pi(f^{\mathcal{A}}(t_1^{\mathcal{A}}(\bar{a}), \ldots, t_s^{\mathcal{A}}(\bar{a}))), \\ &= f^{\mathcal{B}}(\pi(t_1^{\mathcal{A}}(\bar{a})), \ldots, \pi(t_s^{\mathcal{A}}(\bar{a}))) && \text{by definition of an embedding,} \\ &= f^{\mathcal{B}}(t_1^{\mathcal{B}}(\pi(\bar{a})), \ldots, t_s^{\mathcal{B}}(\pi(\bar{a}))) && \text{by the inductive hypothesis,} \\ &= t^{\mathcal{B}}(\pi(\bar{a})), \end{aligned}$$

as required. □

Exercises

2.1 Write out a recursive definition of L_{mon}-terms.

2.2 Explain how the L_{ring}-terms $(-(x + 1) \cdot y)$ and $(x + x) + (y \cdot -z)$ are built up from the recursive definition of L-terms.

2.3 In the structure $\mathbb{R}_{\text{ring}}$, which elements of $\mathbb{R}$ are named by closed L_{ring}-terms?

2.4 Give three different L_{ring}-terms which are interpreted as the same polynomial function.

2.5 Define a polynomial function on $\mathbb{R}$ to be a function $\mathbb{R}^n \to \mathbb{R}$ of the form

$$p(x_1, \ldots, x_n) = \sum_{i=1}^{m} r_i x_1^{d_{i,1}} \cdots x_n^{d_{i,n}}$$

for some $r_i \in \mathbb{R}$ and some $d_{i,j} \in \mathbb{N}$ for $i = 0, \ldots, m$ and $j = 1, \ldots, n$. Show by induction on the construction of terms that every L_{ring}-term is interpreted in $\mathbb{R}_{\text{ring}}$ as a polynomial function. Which polynomial functions $\mathbb{R}^n \to \mathbb{R}$ are the interpretation of some L_{ring}-term?

2.6 Give a complete proof (not using Proposition 2.4) that if $\mathcal{A}$ is an L_{mon}-structure, $\pi \in \text{Aut}(\mathcal{A})$, t is an L_{mon}-term, and $\bar{a} \in A^r$ then $\pi(t^{\mathcal{A}}(\bar{a})) = t^{\mathcal{A}}(\pi(\bar{a}))$.

2.7 Show that if $\mathcal{A} \xrightarrow{\pi} \mathcal{B}$ is an L-homomorphism, then it preserves the interpretations of L-terms.

2.8 An $\mathbb{R}$-linear function is a function of the form $f(x_1, \ldots, x_n) = \sum_{i=1}^{n} r_i x_i$, for some $r_i \in \mathbb{R}$. Design a language $L_{\mathbb{R}\text{-VS}}$ in which the terms can naturally be interpreted as $\mathbb{R}$-linear functions.

2.9 Show how an $\mathbb{R}$-vector space could be interpreted as an $L_{\mathbb{R}\text{-VS}}$-structure, where $L_{\mathbb{R}\text{-VS}}$ is the language of the previous question, and show that an $L_{\mathbb{R}\text{-VS}}$-homomorphism of vector spaces is the same thing as a linear map.

3
Formulas

Terms are certain strings of symbols from the language. As we have seen, they are interpreted as elements of a structure or as functions on the structure. Other strings of symbols of the language say something *about* structures or about elements of the structure. For example, $((1 + 0) + 1) = 0$ is a statement about numbers (which is false in $\mathbb{Z}_{\text{adgp}}$ but true in the cyclic group $\mathbb{Z}/2\mathbb{Z}$), $\exists x[((x \cdot x) + -(1 + 1)) = 0]$ asserts that there is a square root of 2 (which is true in $\mathbb{R}_{\text{ring}}$ but false in $\mathbb{Z}_{\text{ring}}$), and $(x \cdot x) < 1$ is a statement about the number x (which in $\mathbb{R}_{\text{o-ring}}$ will be true for some values of x and false for other values).

3.1 The Recursive Definition of Formulas

Definition 3.1 As for terms, the set of *formulas* of L is defined recursively.

(i) If t_1 and t_2 are terms, then $(t_1 = t_2)$ is a formula.
(ii) If $t_1, \ldots, t_n$ are terms and R is a relation symbol of L of arity n, then $R(t_1, \ldots, t_n)$ is a formula.
(iii) If φ and ψ are formulas, then $\neg\varphi$ and $(\varphi \wedge \psi)$ are formulas.
(iv) If φ is a formula and x is a variable, then $\exists x[\varphi]$ is a formula.
(v) Only something built from the above four clauses in finitely many steps is a formula.

The formulas in clauses (i) and (ii) are called *atomic formulas*, because they do not contain smaller formulas.

3.2 Free Variables and Scope

Variables play two different roles in formulas. For example, the formula

$$x < 1 + 1$$

says something about the variable x, namely that it is less than 2, whereas the formula

$$\exists x[0 < x]$$

says that there is some element which is greater than 0 and the variable x is just a dummy. The formula has the same meaning as $\exists y[0 < y]$, where x is replaced by y. The difference is that in the second example, the variable x is quantified over by the $\exists$ quantifier.

Definition 3.2 An instance of a variable x in a formula is *a quantifier instance* if it occurs immediately after the $\exists$ symbol. Immediately following the quantifier $\exists x$ is the *scope* of the quantifier, which is enclosed in square brackets [ ]. Any instance of x which occurs within the scope of the quantifier $\exists x$ is called a *bound* instance of x, and the quantifier $\exists x$ is said to *bind* it. Any other instance of a variable is said to be a *free* instance. The variables which have free instances in a formula are called its *free variables*.

For simplicity, we will usually assume that no variable occurs both free and bound in the same formula. For example, the formula

$$(x \leqslant y \wedge \exists x[0 \leqslant x])$$

should be rewritten as $(x \leqslant y \wedge \exists z[0 \leqslant z])$.

Formulas with no free variables are particularly important, essentially because they say something about a structure as a whole rather than about some elements of it, so they have a special name.

Definition 3.3 A formula with no free variables is called a *sentence*.

3.3 Interpretation of Formulas

The purpose of formulas of L is to say something about L-structures. A sentence should be either true or false in an L-structure. A formula with free variables may be true for some interpretations of the variables and false for others. We next explain how formulas are interpreted.

Definition 3.4 (Interpretation of formulas) Let φ be a formula of L and $\bar{x} = (x_1, \ldots, x_n)$ a list of variables containing every free variable of φ. We also write $\varphi(\bar{x})$ for the formula with the list of variables. Let $\mathcal{A}$ be an L-structure and $\bar{a} = (a_1, \ldots, a_n)$ a list of elements of (the domain of) $\mathcal{A}$. We define the notion $\mathcal{A} \models \varphi(\bar{a})$, read "$\mathcal{A}$ models $\varphi(\bar{a})$" or "$\varphi(\bar{a})$ is true in $\mathcal{A}$" or "$\mathcal{A}$ satisfies $\varphi(\bar{a})$" by recursion on formulas.

(i) $\mathcal{A} \models t_1(\bar{a}) = t_2(\bar{a})$ iff $t_1^{\mathcal{A}}(\bar{a}) = t_2^{\mathcal{A}}(\bar{a})$, that is, iff the functions $t_1^{\mathcal{A}}$ and $t_2^{\mathcal{A}}$ take the same value at $\bar{a}$.
(ii) $\mathcal{A} \models R(t_1(\bar{a}), \ldots, t_r(\bar{a}))$ iff $(t_1^{\mathcal{A}}(\bar{a}), \ldots, t_r^{\mathcal{A}}(\bar{a})) \in R^{\mathcal{A}}$.
(iii) $\mathcal{A} \models \neg\varphi(\bar{a})$ iff $\mathcal{A} \not\models \varphi(\bar{a})$, that is, $\mathcal{A}$ does not model $\varphi(\bar{a})$.
(iv) $\mathcal{A} \models (\varphi_1 \wedge \varphi_2)(\bar{a})$ iff $\mathcal{A} \models \varphi_1(\bar{a})$ and $\mathcal{A} \models \varphi_2(\bar{a})$.
(v) $\mathcal{A} \models \exists x[\varphi(x, \bar{a})]$ iff there is some $b \in A$ such that $\mathcal{A} \models \varphi(b, \bar{a})$.

Definition 3.5 In the special case where φ is a sentence, if $\mathcal{A} \models \varphi$, we say that $\mathcal{A}$ is a *model* of φ. More generally, if Σ is a set of L-sentences, we say $\mathcal{A}$ is a *model* of Σ and write $\mathcal{A} \models \Sigma$ if $\mathcal{A}$ is a model of every sentence in Σ.

This notion is where the name *model theory* comes from.

The definition above formally gives the formulas, which are strings of symbols, the meaning that we already informally expect them to have. The purpose of the formal definition is that we can use it to prove things that we could not prove just from our informal understanding. Since formulas and terms are defined recursively, the proofs go by induction on the recursive definitions. The following results give good examples of this style of proof.

3.4 Embeddings and Formulas

Lemma 3.6 *An embedding $\mathcal{A} \xrightarrow{\pi} \mathcal{B}$ of L-structures preserves atomic formulas. That is, whenever $\varphi(\bar{x})$ is an atomic L-formula and $\bar{a} \in A^n$,*

$$\mathcal{A} \models \varphi(\bar{a}) \text{ if and only if } \mathcal{B} \models \varphi(\pi(\bar{a})).$$

Proof Let φ be an atomic formula of the form $R(t_1(\bar{x}), \ldots, t_r(\bar{x}))$, and $\bar{a} \in A^n$. Let $\alpha_i = t_i^{\mathcal{A}}(\bar{a})$, let $\beta_i = t_i^{\mathcal{B}}(\pi(\bar{a}))$, and write $\bar{\alpha} = (\alpha_1, \ldots, \alpha_r)$ and $\bar{\beta} = (\beta_1, \ldots, \beta_r)$. Then Proposition 2.4 shows that $\pi(\bar{\alpha}) = \bar{\beta}$. Then, by definition of an embedding, $\bar{\alpha} \in R^{\mathcal{A}}$ iff $\bar{\beta} \in R^{\mathcal{B}}$. Thus $\mathcal{A} \models R(t_1(\bar{a}), \ldots, t_r(\bar{a}))$ iff $\mathcal{B} \models R(t_1(\pi(\bar{a})), \ldots, t_r(\pi(\bar{a})))$, that is, $\mathcal{A} \models \varphi(\bar{a})$ iff $\mathcal{B} \models \varphi(\pi(\bar{a}))$ as required.

The proof is similar in the case when φ is an atomic formula of the form $t_1(\bar{x}) = t_2(\bar{x})$. □

Proposition 3.7 *An isomorphism $\mathcal{A} \xrightarrow{\pi} \mathcal{B}$ of L-structures preserves every L-formula. That is, whenever $\varphi(\bar{x})$ is an L-formula and $\bar{a} \in A^n$,*

$$\mathcal{A} \models \varphi(\bar{a}) \text{ if and only if } \mathcal{B} \models \varphi(\pi(\bar{a})).$$

Proof We proceed by induction on the construction of formulas. If φ is an atomic formula, then the result is given by the preceding lemma.

Suppose the result is true for φ and ψ.

$$
\begin{aligned}
\neg \text{ case:} \quad \mathcal{A} \models \neg\varphi(\bar{a}) \quad &\text{iff} \quad \mathcal{A} \not\models \varphi(\bar{a}),\\
&\text{iff} \quad \mathcal{B} \not\models \varphi(\pi(\bar{a})) \qquad \text{by induction hypothesis,}\\
&\text{iff} \quad \mathcal{B} \models \neg\varphi(\pi(\bar{a})).
\end{aligned}
$$

$$
\begin{aligned}
\wedge \text{ case:} \quad \mathcal{A} \models (\varphi \wedge \psi)(\bar{a}) \quad &\text{iff} \quad \mathcal{A} \models \varphi(\bar{a}) \text{ and } \mathcal{A} \models \psi(\bar{a}),\\
&\text{iff} \quad \mathcal{B} \models \varphi(\pi(\bar{a})) \text{ and}\\
&\qquad \mathcal{B} \models \psi(\pi(\bar{a})) \qquad \text{by induction hypothesis,}\\
&\text{iff} \quad \mathcal{B} \models (\varphi \wedge \psi)(\pi(\bar{a})).
\end{aligned}
$$

$\exists x$ case: Suppose $\mathcal{A} \models \exists x\varphi(x, \bar{a})$. Then there is $c \in A$ such that $\mathcal{A} \models \varphi(c, \bar{a})$. By induction, $\mathcal{B} \models \varphi(\pi(c), \pi(\bar{a}))$. So $\mathcal{B} \models \exists x\varphi(x, \pi(\bar{a}))$. Conversely, suppose $\mathcal{B} \models \exists x\varphi(x, \pi(\bar{a}))$. Then there is $d \in B$ such that $\mathcal{B} \models \varphi(d, \pi(\bar{a}))$. Since π is an isomorphism, it has an inverse, π^{-1}. Let $c = \pi^{-1}(d)$. Then $\pi(c) = d$, so, by induction, $\mathcal{A} \models \varphi(c, \bar{a})$. So $\mathcal{A} \models \exists x\varphi(x, \bar{a})$.

That completes all the necessary induction steps, one for each part of the recursive definition of a formula. □

3.5 Abbreviations

Our formal language is very restricted. In practice, we adopt many abbreviations which make it easier to use. We write

not equal	$(t_1 \neq t_2)$	for	$\neg(t_1 = t_2)$,
or	$(\varphi \vee \psi)$	for	$\neg(\neg\varphi \wedge \neg\psi)$,
implies	$(\varphi \to \psi)$	for	$(\neg\varphi \vee \psi)$,
iff	$(\varphi \leftrightarrow \psi)$	for	$((\varphi \to \psi) \wedge (\psi \to \varphi))$,
for all	$\forall x[\varphi]$	for	$\neg\exists x[\neg\varphi]$.

We could have taken all these symbols as part of our formal language, but if we did that, then our proofs about the formal language would be longer, so instead we introduce them as abbreviations.

We will omit brackets or sometimes add extra brackets where this improves clarity and where no ambiguity can arise. We will also write $\exists x_1 \cdots x_n[\varphi]$ for $\exists x_1[\exists x_2[\cdots \exists x_n[\varphi] \cdots]]$ and $\bigwedge_{i=1}^{n} \varphi_i$ for $\varphi_1 \wedge \varphi_2 \wedge \cdots \wedge \varphi_n$. We adopt the usual convention of writing $(x + y)$ instead of $+(x, y)$, and similarly other standard conventions from algebra. Our general philosophy is to use the usual notation from mathematical practice where possible, but be careful to see that it can, in principle, be translated into our formal language.

Exercises

3.1 Write down an L_{ring}-formula in which the variables x, y, and z appear free and the variables u and v are bound.

3.2 In the language L with a binary relation symbol $<$ and constant symbols for $0, 1, 2, 3, 4, 5, 6$, write L-sentences expressing the following about the L-structure on the integers, $\mathbb{Z}$.

(a) There is no greatest integer.
(b) There is no integer between 0 and 1.
(c) For any two distinct integers, one is less than the other.
(d) There are exactly 2 integers between 3 and 6.

3.3 Let $L_=$ be the language with no relation, function, or constant symbols. For $n \in \mathbb{N}^+$, show there are $L_=$-sentences $\varphi_{\geqslant n}$ and $\varphi_{=n}$ such that for any structure $\mathcal{A}$ we have $\mathcal{A} \models \varphi_{\geqslant n}$ iff A has at least n elements and $\mathcal{A} \models \varphi_{=n}$ iff A has exactly n elements.

3.4 Consider the structure $\mathcal{N} = \langle \mathbb{N}; +, \cdot \rangle$. Write down

(a) a formula $\varphi(x)$ in the language $\langle +, \cdot \rangle$ with one free variable such that for any $n \in \mathbb{N}$, $\mathcal{N} \models \varphi(n)$ iff $n = 2$;
(b) a formula $\xi(x, y)$ with two free variables such that for any $n_1, n_2 \in \mathbb{N}$, $\mathcal{N} \models \xi(n_1, n_2)$ iff $n_1 < n_2$;
(c) a formula $\delta(x, y)$ with two free variables such that for any $n_1, n_2 \in \mathbb{N}$, $\mathcal{N} \models \delta(n_1, n_2)$ iff $n_1 | n_2$;
(d) a formula $\psi(x)$ with one free variable such that for any $n \in \mathbb{N}$, $\mathcal{N} \models \psi(n)$ iff n is a prime number;
(e) a sentence stating that there are infinitely many pairs of prime numbers which differ by 2.

You can use abbreviations, provided you explain what they are.

3.5 Consider the L_{ring}-formula $\varphi(x)$ given by $\exists y[(y \cdot y) + 1 = x]$. For what values of x do we have $\mathbb{R}_{\text{ring}} \models \varphi(x)$? For what x do we have $\mathbb{Z}_{\text{ring}} \models \varphi(x)$?

3.6 Express each of the statements in Exercise 1.2 as L-sentences for an appropriate choice of language L.

3.7 A formula is said to be *quantifier-free* if it does not use the symbol $\exists$ (or the abbreviation $\forall$).

(a) Write down a recursive definition of quantifier-free formulas in the style of Definition 3.1.

(b) Prove by induction on your recursive definition that every embedding $\mathcal{A} \xrightarrow{\pi} \mathcal{B}$ of L-structures preserves quantifier-free formulas. That is, whenever $\varphi(\bar{x})$ is a quantifier-free L-formula and $\bar{a} \in A^n$,

$$\mathcal{A} \models \varphi(\bar{a}) \text{ if and only if } \mathcal{B} \models \varphi(\pi(\bar{a})).$$

3.8 A *positive quantifier-free formula* is one defined using $\wedge$, $\vee$, and atomic formulas, but not $\neg$, $\rightarrow$, or quantifiers. Write down a recursive definition of positive quantifier-free formulas, and prove that they are preserved under L-homomorphisms.

4
Definable Sets

In this chapter we introduce the main idea of the book, that of a definable set.

4.1 Examples of Definable Sets

Definition 4.1 In an L-structure $\mathcal{A}$, a subset S of A^n is said to be a *definable set* if there is an L-formula $\varphi(x_1, \ldots, x_n)$ such that $S = \{\bar{a} \in A^n \mid \mathcal{A} \models \varphi(\bar{a})\}$. The formula $\varphi(\bar{x})$ *defines* the set S, and we sometimes write S as $\varphi(\mathcal{A})$.

Example 4.2 In $\mathbb{R}_{\text{o-ring}}$, the interval $(0, 2)$ is defined by the formula

$$(0 < x) \wedge (x < (1 + 1)),$$

the pair $\{-1, 3\}$ is defined by the formula

$$(x + 1 = 0) \vee (x = (1 + 1) + 1),$$

and the formula

$$(0 = x \vee 0 < x) \wedge (x < 1 + 1) \wedge (y < 0) \wedge (-(1 + 1) < y)$$

defines the rectangle $[0, 2) \times (-2, 0)$.

If $t(x_1, \ldots, x_n)$ is an L-term, then the formula $y = t(x_1, \ldots, x_n)$ defines the graph of the function $t^{\mathcal{A}}$ in the structure $\mathcal{A}$. However, other functions may also be definable.

Definition 4.3 Let $\mathcal{A}$ be an L-structure, $n, m \in \mathbb{N}^+$, $D \subseteq A^n$, $C \subseteq A^m$, and $f : D \to C$ a function. Then f is a *definable function* (with respect to the L-structure $\mathcal{A}$) if D, C and the graph of f, that is $\{(a, f(a)) \mid a \in D\}$, are definable sets.

Note that if the graph of f is defined by the formula $\varphi(x, y)$, then the domain of f is defined by $\exists y[\varphi(x, y)]$ and the image is defined by $\exists x[\varphi(x, y)]$.

Example 4.4 In $\mathbb{R}_{\text{o-ring}}$, the absolute value function $\mathbb{R} \to \mathbb{R}; x \mapsto |x|$ is defined by the formula

$$((x < 0) \wedge (y = -x)) \vee (\neg(x < 0) \wedge (y = x)).$$

Lemma 4.5 *If $S_1, S_2 \subseteq A^n$ are definable, then $S_1 \cap S_2$, $S_1 \cup S_2$, and $A^n \setminus S_1$ are definable. Furthermore, if $S \subseteq A^{n+m}$ is definable, then the projection*

$$\operatorname{proj}(S) = \left\{\bar{a} \in A^n \,\middle|\, \textit{There is some } \bar{b} \in A^m \textit{ such that } (\bar{a}, \bar{b}) \in S\right\}$$

is also definable.

Proof Suppose S_1 is defined by $\varphi(\bar{x})$ and S_2 is defined by $\psi(\bar{x})$. Then $S_1 \cap S_2$ is defined by $(\varphi(\bar{x}) \wedge \psi(\bar{x}))$ and so on. □

Figure 4.1 illustrates why existential quantification corresponds to projection. In the structure $\mathbb{R}_{\text{o-ring}}$, let $\varphi(x, y)$ be the formula

$$(x - 4)^2 + (y - 3)^2 < 4$$

which defines a disc of radius 2, centred at $(4, 3)$ in the xy-plane. We have used subtraction and squaring in the formula, although they are not symbols in the language $L_{\text{o-ring}}$. Since they are definable functions, we can use them as abbreviations. The formula $\exists y[\varphi(x, y)]$ defines the projection to the x-axis, which is the interval $(2, 6)$.

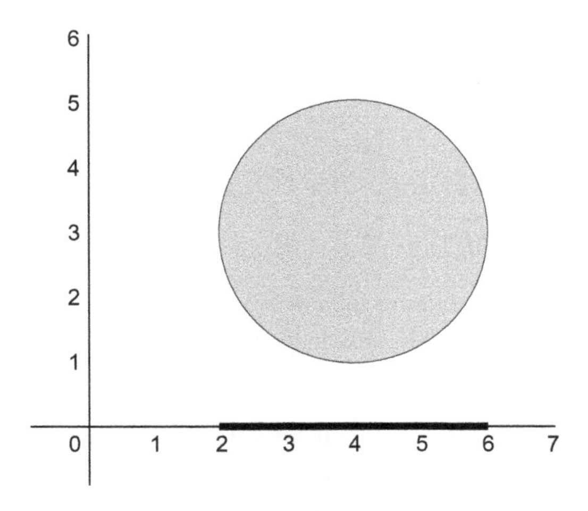

Figure 4.1 Existential quantification as projection.

The subsets of $\mathbb{R}^n$ which are defined by $L_{\text{o-ring}}$-formulas without quantifiers, that is, by equations and inequations of polynomials, are called *semi-algebraic sets*. In fact, for the structure $\mathbb{R}_{\text{o-ring}}$, using quantifiers does not give any more definable sets. So, for example, one can prove that $\mathbb{Z}$ is not a definable subset of $\mathbb{R}$ in $\mathbb{R}_{\text{o-ring}}$. See Proposition 22.3 for more detail, although a complete proof is beyond the scope of this book.

4.2 Preservation of Definable Sets under Automorphisms

A very important part of model theory is the study of the definable sets of a structure $\mathcal{A}$. To show a set is definable, we can just come up with a formula, but it is more difficult to show that no formula defines a set. So it is important to have methods to show that certain subsets are *not* definable. We now develop one such method.

Proposition 4.6 *If $S \subseteq A^n$ is a definable set and π is an automorphism of $\mathcal{A}$, then $\pi(S) = S$. We say that S is* preserved *under all automorphisms.*

Proof By $\pi(S)$, we mean $\{\pi(\bar{a}) \mid \bar{a} \in S\}$, the image of S under the function π. Suppose S is defined by the formula $\varphi(\bar{x})$, and let $\bar{a} \in S$. Then $\mathcal{A} \models \varphi(\bar{a})$, so applying Proposition 3.7, we see that $\mathcal{A} \models \varphi(\pi(\bar{a}))$ because π is an automorphism, and hence $\pi(\bar{a}) \in S$. So $\pi(S) \subseteq S$. Now π^{-1} is also an automorphism of $\mathcal{A}$, so $\pi^{-1}(\bar{a}) \in S$. But $\bar{a} = \pi(\pi^{-1}(\bar{a}))$, so $\bar{a} \in \pi(S)$. Thus $S \subseteq \pi(S)$, and hence $\pi(S) = S$, as required. □

Example 4.7 In the structure $\mathbb{R}_{\text{adgp}}$, the order $<$ is not definable. That is, the set of pairs $\left\{(x, y) \in \mathbb{R}^2 \mid x < y\right\}$ is not definable. To see this, consider the function $\pi : \mathbb{R} \to \mathbb{R}$ given by $\pi(x) = -x$. It is an embedding of $\mathbb{R}_{\text{adgp}}$ into itself because $\pi(0) = 0$ and

$$\pi(x + y) = -(x + y) = -x + (-y) = \pi(x) + \pi(y).$$

Furthermore, $\pi \circ \pi$ is the identity map on $\mathbb{R}$, so π is its own inverse. Hence π is an automorphism of $\mathbb{R}_{\text{adgp}}$. However, we have for example $0 < 1$, but $\pi(0) \not< \pi(1)$, and so $<$ is not preserved under π, and hence by Proposition 4.6 it is not definable in $\mathbb{R}_{\text{adgp}}$.

It is natural to ask if the converse to Proposition 4.6 is true. If $S \subseteq A^n$ is a subset which is not definable, must there be an automorphism π of $\mathcal{A}$ such that $\pi(S) \neq S$? The answer is no.

Lemma 4.8 *There are no non-trivial automophisms of* $\mathbb{R}_{\text{o-ring}}$.

Proof Suppose π is an automorphism of $\mathbb{R}_{\text{o-ring}}$. Then $\pi(0) = 0$. Furthermore, for any $n \in \mathbb{N}^+$, we have $\pi(\underbrace{1 + \cdots + 1}_{n}) = \underbrace{\pi(1) + \cdots + \pi(1)}_{n} = n$, and then $\pi(-n) = -\pi(n) = -n$, so π fixes every integer. If $q = n/m \in \mathbb{Q}$, we have $\pi(q) \cdot \pi(m) = \pi(n)$, so $\pi(q) \cdot m = n$, but q is the only solution to the equation $x \cdot m = n$, which is an $L_{\text{o-ring}}$-formula, so we must have $\pi(q) = q$. So π fixes every rational number. The rationals are dense in the reals and π fixes the order $<$, so it must fix every real number. Hence π is the identity map on $\mathbb{R}$. □

There are uncountably many subsets of $\mathbb{R}^n$, and only countably many formulas, so only countably many definable subsets, so most of the subsets of $\mathbb{R}^n$ are not definable. However, all subsets of $\mathbb{R}^n$ are preserved by the identity automorphism and hence by all automorphisms of $\mathbb{R}_{\text{o-ring}}$.

Exercises

4.1 Complete the proof of Lemma 4.5.

4.2 Show that addition and multiplication are not definable in $\mathbb{R}_<$.

4.3 Show that the order $<$ is not definable in $\mathbb{Z}_{\text{adgp}}$ but is definable in $\mathbb{N}_{\text{admon}}$.

4.4 Show that the order $<$ is definable in $\mathbb{R}_{\text{ring}}$. Deduce that every subset of $\mathbb{R}^n$ which is definable in $\mathbb{R}_{\text{o-ring}}$ is also definable in $\mathbb{R}_{\text{ring}}$.

4.5 Show that the set $\mathbb{N}$ and the order $<$ are definable in $\mathbb{Z}_{\text{ring}}$. [Hint: Use Lagrange's 4-square theorem.]

4.6 Show that negation is a definable function in $\langle \mathbb{R}; +, 0\rangle$ and in $\langle \mathbb{Z}; +, 0\rangle$ but that multiplication is not definable in either structure.

4.7 Show that the square root function $\mathbb{R}^+ \to \mathbb{R}^+; x \mapsto \sqrt{x}$ is definable in $\mathbb{R}_{\text{o-ring}}$.

4.8 Give an example of a definable function in $\mathbb{Z}_{\text{ring}}$ which is not defined by the interpretation of any term.

4.9 Show that the only automorphism of the structure $\mathbb{N}_{\text{s-ring}}$ is the identity map.

4.10 Consider the expansion of the real ordered field $\mathbb{R}_{\text{o-ring}}$ by a unary function f. Write down a formula $\varphi(x)$ such that whatever the function f is, $\varphi(x)$ defines the set of real numbers at which f is continuous.

4.11 Find all the subsets of $\mathbb{Q}$ which are definable in $\mathbb{Q}_{\text{adgp}}$ (by a formula with one free variable) and prove that they are the only ones.

4.12 Sketch a proof of Proposition 4.6 starting from the relevant definitions and including the statements and sketch proofs of any necessary lemmas. You should explain what methods of proof are used, what all the key ideas are, and how they fit together, but you do not need to give all the details of the inductive proofs.

5
Substructures and Quantifiers

In the previous chapter we showed that every definable set is preserved under automorphisms. Here we give more subtle preservation theorems which show that formulas constructed with only certain quantifiers are preserved under taking substructures or embeddings.

5.1 Substructures

The field of rational numbers, $\mathbb{Q}_{\text{ring}}$, is a *subfield* of the field of real numbers, $\mathbb{R}_{\text{ring}}$. Similarly, $\mathbb{Z}_{\text{ring}}$ is a *subring* of $\mathbb{Q}_{\text{ring}}$. More generally, recall from Chapter 1 the notion of a *substructure* of an L-structure.

Definition 5.1 Suppose that $\mathcal{A}$ and $\mathcal{B}$ are L-structures. Then $\mathcal{A}$ is a *substructure* of $\mathcal{B}$ and $\mathcal{B}$ is an *extension* of $\mathcal{A}$ if the domain of $\mathcal{A}$ is a subset of the domain of $\mathcal{B}$ and all the symbols of L are interpreted in $\mathcal{A}$ as the restrictions of their interpretations in $\mathcal{B}$. More precisely:

- for each constant symbol c, $c^{\mathcal{A}} = c^{\mathcal{B}}$;
- for each function symbol f and each $\bar{a}$ in A, $f^{\mathcal{A}}(\bar{a}) = f^{\mathcal{B}}(\bar{a})$; and
- for each relation symbol R and each $\bar{a}$ in A, $\bar{a} \in R^{\mathcal{A}}$ if and only if $\bar{a} \in R^{\mathcal{B}}$.

Equivalently, A is a subset of B, and the inclusion map of A into B is an L-embedding. We write $\mathcal{A} \subseteq \mathcal{B}$ to mean that $\mathcal{A}$ is an L-substructure of $\mathcal{B}$.

5.2 Existential and Universal Formulas

For $\mathcal{A} \subseteq \mathcal{B}$, we now consider how the interpretations of formulas are related in the two structures. Consider the following sentences, at first written in English sentences, not as L-sentences, for a first-order language L.

(i) 'All numbers are greater than or equal to 0.'
(ii) 'There is a number which when added to itself makes 1.'
(iii) 'All numbers have square roots.'

The first sentence is true when we interpret "number" as an element of the structure $\langle \mathbb{N}; <, 0\rangle$, and it is also true in the substructure $\langle E; <, 0\rangle$ consisting of the even natural numbers, but it is false in the extension $\langle \mathbb{Z}; <, 0\rangle$. Intuitively, it makes sense, because if all natural numbers have a property, then certainly all the even natural numbers have that property, but perhaps some numbers in a larger set (the integers) might not have that property. For $L = \langle 0, <\rangle$, we can write the sentence as the L-sentence

$$\forall x[0 < x \vee x = 0]$$

with a universal quantifier. The second sentence is true in $\mathbb{Q}_{\text{ring}}$ and also in the extension $\mathbb{R}_{\text{ring}}$ but false in the substructure $\mathbb{Z}_{\text{ring}}$. It makes sense, because once the number (1/2) with this property is there in $\mathbb{Q}$, it is still in any extension of $\mathbb{Q}$, but it might not be there in a substructure, in this case, $\mathbb{Z}$. We can write it as

$$\exists x[x + x = 1]$$

using an existential quantifier.

The third sentence is true in the complex field $\mathbb{C}_{\text{ring}}$. It appears to be another universal statement since it is about 'all numbers', but it is false in the substructure $\mathbb{R}_{\text{ring}}$. It is also false in the ring $\mathbb{C}[X]$ of polynomials over $\mathbb{C}$, which is an extension ring of $\mathbb{C}$. In fact, when we write it as the L_{ring}-sentence $\forall x \exists y[y \cdot y = x]$, we see that it uses both universal and existential quantifiers.

The first two sentences are examples of a general phenomenon, that universal statements true in a structure are also true in any substructure, and that existential statements true in a structure are also true in any extension. The pattern also holds for formulas with free variables. In our convention, universal quantifiers are just abbreviations and can be replaced by existential quantifiers, but only by also using negation symbols. So we need to be very careful about exactly what we mean by universal and existential statements.

Definition 5.2 (Quantifier-free formula) A *quantifier-free formula* is a formula which does not involve any quantifiers. In order to prove things about them, it is useful also to give a recursive definition.

(i) An atomic formula is a quantifier-free formula.
(ii) If φ is a quantifier-free formula, then $\neg\varphi$ is a quantifier-free formula.
(iii) If φ and ψ are quantifier-free formulas, then $(\varphi \wedge \psi)$ is a quantifier-free formula.

(iv) Only a formula constructed by the above rules in finitely many steps is a quantifier-free formula.

Similarly, we can give recursive definitions of existential formulas and universal formulas.

Definition 5.3 (Existential formula)

(i) A quantifier-free formula is an existential formula.
(ii) If φ is an existential formula and x is a variable, then $\exists x[\varphi]$ is an existential formula.
(iii) Only a formula constructed by the above rules in finitely many steps is an existential formula.

Definition 5.4 (Universal formula)

(i) A quantifier-free formula is a universal formula.
(ii) If φ is a universal formula and x is a variable, then $\forall x[\varphi]$ is a universal formula.
(iii) Only a formula constructed by the above rules in finitely many steps is a universal formula.

So any existential formula has the form $\exists y_1 \cdots y_n[\psi]$ where ψ is a quantifier-free formula, and any universal formula has the form $\forall y_1 \cdots y_n[\psi]$. From our definition of $\forall y$ as an abbreviation for $\neg\exists y\neg$, we can write $\forall y_1 \cdots y_n[\psi]$ as $\neg\exists y_1\neg\neg\exists y_2\neg \cdots \neg\exists y_n[\neg\psi]$ and cancelling the double negations it becomes $\neg\exists y_1 \cdots y_n[\neg\psi]$, which is the negation of an existential formula.

5.3 Preservation Laws

Now we can prove that existential statements are indeed 'preserved in extensions' and universal statements are 'preserved in substructures'.

Lemma 5.5 *If $\mathcal{A}$ is an L-substructure of $\mathcal{B}$, $\varphi(\bar{x})$ is a quantifier-free L-formula, and $\bar{a} \in A$, then*

$$\mathcal{A} \models \varphi(\bar{a}) \text{ if and only if } \mathcal{B} \models \varphi(\bar{a}).$$

Proof If $\varphi(\bar{x})$ is an atomic L-formula, the result is immediate from Lemma 3.6. More generally, it is Exercise 3.6. □

Proposition 5.6 *Let $\mathcal{B}$ be an L-structure and $\mathcal{A} \subseteq \mathcal{B}$ an L-substructure. Let $\varphi(\bar{x})$ be an existential L-formula, and suppose $\bar{a} \in A^n$ is such that we have $\mathcal{A} \models \varphi(\bar{a})$. Then $\mathcal{B} \models \varphi(\bar{a})$.*

Proof We proceed by induction on the construction of existential formulas. If $\varphi(\bar{x})$ is a quantifier-free formula, then the result is Lemma 5.5. Suppose $\varphi(\bar{x})$ is $\exists y[\psi(\bar{x}, y)]$. By assumption, $\mathcal{A} \models \exists y[\psi(\bar{a}, y)]$, so there is $b \in A$ such that $\mathcal{A} \models \psi(\bar{a}, b)$. Then, by induction hypothesis, $\mathcal{B} \models \psi(\bar{a}, b)$. So $\mathcal{B} \models \exists y[\psi(\bar{a}, y)]$, as required. □

Proposition 5.7 *Let $\mathcal{B}$ be an L-structure and $\mathcal{A} \subseteq \mathcal{B}$ an L-substructure. Let $\varphi(\bar{x})$ be a universal L-formula, and suppose $\bar{a} \in A$ is such that $\mathcal{B} \models \varphi(\bar{a})$. Then $\mathcal{A} \models \varphi(\bar{a})$.*

Proof We could do a simple inductive proof as for the existential case, but instead we will deduce it from that case. Let θ be an existential formula such that φ is $\neg\theta$. Suppose that $\mathcal{B} \models \varphi(\bar{a})$. Then $\mathcal{B} \not\models \theta(\bar{a})$, so, by Proposition 5.6, $\mathcal{A} \not\models \theta(\bar{a})$. So $\mathcal{A} \models \varphi(\bar{a})$, as required. □

Exercises

5.1 If L is a language with only relation symbols and $\mathcal{A}$ is an L-structure, show that every subset of A is naturally an L-substructure of $\mathcal{A}$.

5.2 If L has function and constant symbols, which subsets of the domain of an L-structure correspond to L-substructures?

5.3 Show that an L_{ring}-substructure of a field is a subring. Design a language L_{field} for fields such that L_{field}-substructures of fields are subfields.

5.4 Write the axiom for a group which states that every element has an inverse both as an L_{gp}-sentence and as an L_{mon}-sentence. Which of the axioms is a universal sentence? Show that if G is a group considered as an L_{gp}-structure, then its L_{gp}-substructures are its subgroups and its L_{mon}-substructures are submonoids.

5.5 Show that every subgroup of an abelian group is also abelian.

5.6 Show that there is no universal L_{ring}-sentence which asserts that every non-zero element of a field has a multiplicative inverse.

5.7 Give a direct proof of Proposition 5.7, not using Proposition 5.6.

5.8 Suppose that $\mathcal{A}$ and $\mathcal{B}$ are L-structures and $\mathcal{A} \subseteq \mathcal{B}$. Let $\varphi(x_1, \ldots, x_n)$ be an L-formula. Show that:

(a) if φ is quantifier-free, then $\varphi(\mathcal{A}) = \varphi(\mathcal{B}) \cap A^n$;
(b) if φ is existential, then $\varphi(\mathcal{A}) \subseteq \varphi(\mathcal{B}) \cap A^n$; and
(c) if φ is universal, then $\varphi(\mathcal{A}) \supseteq \varphi(\mathcal{B}) \cap A^n$.

5.9 Suppose we have a chain of L-structures

$$\mathcal{A}_1 \subseteq \mathcal{A}_2 \subseteq \mathcal{A}_3 \subseteq \cdots \subseteq \mathcal{A}_n \subseteq \cdots$$

indexed by $n \in \mathbb{N}^+$, and $\mathcal{A}$ is the union of the chain. Explain how $\mathcal{A}$ can be made into an L-structure which is an extension of each $\mathcal{A}_n$. Let $\varphi(\bar{x})$ be a $\forall\exists$-formula, that is, an L-formula of the form $\forall\bar{y}\exists\bar{z}\psi(\bar{x}, \bar{y}, \bar{z})$, with ψ quantifier-free. Show that if $\bar{a} \in A_1$ such that for every $n \in \mathbb{N}^+$ we have $\mathcal{A}_n \models \varphi(\bar{a})$, then $\mathcal{A} \models \varphi(\bar{a})$.

5.10 A *positive L-formula* is defined using $\wedge$, $\vee$, $\exists$, and $\forall$, but without negations (except implicitly in the abbreviations $\vee$ and $\forall$).

(a) Give a recursive definition of positive L-formulas.
(b) Show that positive L-formulas are preserved under surjective homomorphisms of L-structures.
(c) What formulas are preserved under arbitrary (not necessarily surjective) homomorphisms?

Part II

Theories and Compactness

In this part of the book we introduce the first part of our programme for understanding a mathematical structure: finding axioms for the theory of the structure. The programme is illustrated with the Peano axioms for $\mathbb{N}_{\text{s-ring}}$ and by the axioms for algebraically closed fields and for real-closed fields for $\mathbb{C}_{\text{ring}}$ and $\mathbb{R}_{\text{o-ring}}$, respectively. At this stage, we lack the tools to give proofs for these examples, or indeed for any other examples. We then introduce the most powerful of the model-theoretic tools, the compactness theorem, together with the method of changing the language by adding new constant symbols. These methods are used to construct new models of the theory of an infinite structure. Further applications of the compactness theorem are given in the context of axiomatisable classes. A brief discussion of cardinal arithmetic and Henkin's proof of the compactness theorem complete Part II.

6
Theories and Axioms

In this chapter we move the focus from individual formulas and sentences to the set of al those sentences which are true in a structure, its *theory*. To describe a theory, we need to have a subset of it we can write down, that is, an axiomatisation.

6.1 Theories

Recall from Definition 3.5 that if $\mathcal{A}$ is an L-structure and Σ is a set of L-sentences, we write $\mathcal{A} \models \Sigma$, read '$\mathcal{A}$ models Σ', to mean that every sentence σ in Σ is true in $\mathcal{A}$.

Definition 6.1 Let C be a class of L-structures. The *theory* of C, $\text{Th}(C)$, is the set of all L-sentences which are true in every $\mathcal{A} \in C$. When $C = \{\mathcal{A}\}$, a single structure, we write $\text{Th}(\mathcal{A})$ for $\text{Th}(\{\mathcal{A}\})$.

For any class of structures C, the theory $\text{Th}(C)$ has the property of being *deductively closed*, which we now explain.

Definition 6.2 (Entailment) Let Σ be a set of L-sentences and φ be an L-sentence. We say Σ *entails* φ or φ *is a logical consequence of* Σ and write $\Sigma \vdash \varphi$ if every model of Σ is also a model of φ. If Φ is also a set of sentences, we write $\Sigma \vdash \Phi$ to mean that for all $\varphi \in \Phi$, $\Sigma \vdash \varphi$.

Often, the symbol $\models$ is used instead of $\vdash$ in this context, as well as when a structure is on the left-hand side. Then the symbol $\vdash$ is used instead only for a syntactic concept of provability within a given deduction system. Since we do not use the concept of provability, we will follow Wilfrid Hodges [Hod93] and use the symbols this way.

Definition 6.3 A set Σ of L-sentences is *deductively closed* if, for any L-sentence σ, if $\Sigma \vdash \sigma$ then $\sigma \in \Sigma$. The *deductive closure* of Σ is $\{\sigma \mid \Sigma \vdash \sigma\}$.

A set Σ is *satisfiable* if there is an L-structure $\mathcal{A}$ such that $\mathcal{A} \models \Sigma$, that is, if Σ has a model, and is *unsatisfiable* otherwise.

Lemma 6.4 *For any class C of structures,* $\mathrm{Th}(C)$ *is deductively closed.*

Proof Suppose σ is an L-sentence and $\mathrm{Th}(C) \vdash \sigma$. If $\mathcal{A} \in C$, then $\mathcal{A} \models \mathrm{Th}(C)$, so by the definition of $\vdash$, we have $\mathcal{A} \models \sigma$. Hence $\sigma \in \mathrm{Th}(C)$. So $\mathrm{Th}(C)$ is deductively closed. □

Definition 6.5 (Theories) A *theory* (more precisely, a *first-order L-theory*) is a satisfiable and deductively closed set of L-sentences.

Some authors use the word *theory* to mean any set of sentences, and others require satisfiability but not deductive closedness, or vice versa. In practice it rarely matters. A minor quirk of our convention is that if C is the empty set of L-structures, then $\mathrm{Th}(C)$ is the set of all L-sentences. So in this case (and only in this case), $\mathrm{Th}(C)$ is not satisfiable and so is not a theory.

6.2 Examples of Axioms

In practice (and often even in principle) we cannot write down or even describe all the sentences in $\mathrm{Th}(C)$. Instead we want to have a set of *axioms* for C, which means a set of sentences $\Sigma \subseteq \mathrm{Th}(C)$ such that an L-structure $\mathcal{A}$ is in C iff $\mathcal{A} \models \Sigma$. To be useful, we should be able to write these axioms down, or at least describe them. Often a class of structures is defined by a list of axioms, and for illustration, we give some familiar algebraic examples.

Example 6.6 (Axioms for the theory of groups) We can write the following axioms for groups in the language $L_{\mathrm{gp}} = \langle \cdot, (-)^{-1}, 1 \rangle$:

G1. $\forall xyz[(x \cdot y) \cdot z = x \cdot (y \cdot z)]$,
G2. $\forall x[x \cdot 1 = x \wedge 1 \cdot x = x]$, and
G3. $\forall x[x \cdot x^{-1} = 1 \wedge x^{-1} \cdot x = 1]$.

The *theory of groups* is the *deductive closure* of these three axioms. In other words, it is the set of all L_{gp}-sentences which are true of all groups. So, for example, the sentence

$$\forall xy[x^{-1} \cdot (x \cdot y) = y]$$

follows easily from the axioms so is in the theory of groups, but the sentence

$$\exists x[x \neq 1 \wedge x \cdot x = 1]$$

is true in some groups (for example C_2, the cyclic group of order 2), but not in other groups (for example C_3, the cyclic group of order 3), so is not in the theory of groups.

Example 6.7 (Abelian groups) A group is *abelian* if it satisfies also

AG. $\forall xy[x \cdot y = y \cdot x]$,

so the theory of abelian groups is the deductive closure of $\{G1, G2, G3, AG\}$.

We have written four axioms for abelian groups, but we could write a single axiom with the same meaning, the conjunction of those four axioms:

$$G1 \wedge G2 \wedge G3 \wedge AG.$$

The natural language in which to axiomatise rings or fields is $L_{\text{ring}} = \langle +, \cdot, -, 0, 1 \rangle$, the language of rings.

Example 6.8 (Rings) We have the axioms for abelian groups, but now in the additive language $L_{\text{adgp}} = \langle +, -, 0 \rangle$:

R1. $\forall xyz[(x + y) + z = x + (y + z)]$,
R2. $\forall x[x + 0 = x \wedge 0 + x = x]$,
R3. $\forall x[x + (-x) = 0 \wedge (-x) + x = 0]$, and
R4. $\forall xy[x + y = y + x]$.

Then there are axioms for commutative monoids (the abelian group axioms except for the inverse) in the language $L_{\text{mon}} = \langle \cdot, 1 \rangle$:

R5. $\forall xyz[(x \cdot y) \cdot z = x \cdot (y \cdot z)]$,
R6. $\forall x[x \cdot 1 = x \wedge 1 \cdot x = x]$,
R7. $\forall xy[(x \cdot y) = (y \cdot x)]$,

and a distributivity axiom;

R8. $\forall xyz[x \cdot (y + z) = (x \cdot y) + (x \cdot z)]$.

Together these eight sentences axiomatise the theory of (commutative) rings.

Example 6.9 (Fields) A field is a ring satisfying two additional axioms. First, an axiom to rule out the zero ring,

F1. $0 \neq 1$,

and second an axiom which says that every non-zero element of the field has a multiplicative inverse:

F2. $\forall x[x = 0 \vee \exists y[x \cdot y = 1]]$.

So the axioms R1–R8, F1, and F2 together axiomatise the theory of fields.

Remark 6.10 Notice that we put $-$, the additive inverse, as a function symbol in the language. However, we cannot easily do the same for the multiplicative inverse, because it is not defined at 0. We could put a unary function symbol for it in the language, but we would have to define 0^{-1} somehow arbitrarily, and then make sure that our axioms describing how $(-)^{-1}$ works in fields took this case into account. We follow the usual practice of not introducing a symbol for it.

6.3 Peano Arithmetic and Complete Theories

One of the most important structures in mathematics is $\mathbb{N}_{\text{s-ring}} = \langle \mathbb{N}; +, \cdot, 0, 1 \rangle$, the natural numbers with both addition and multiplication. In mathematical logic, the study of this structure and its theory (and related theories) is often called *arithmetic*, and $\text{Th}(\mathbb{N}_{\text{s-ring}})$ is sometimes called *true arithmetic*. We can try to write down axioms for $\text{Th}(\mathbb{N}_{\text{s-ring}})$. The most common and useful axioms are those written down by the Italian mathematician Peano. In modern form, they are as follows:

P1. $\forall x[x + 1 \neq 0]$,
P2. $\forall x[x \neq 0 \to \exists y[x = y + 1]]$,
P3. $\forall xy[x + 1 = y + 1 \to x = y]$,
P4. $\forall x[x + 0 = x]$,
P5. $\forall xy[x + (y + 1) = (x + y) + 1]$,
P6. $\forall x[x \cdot 0 = 0]$,
P7. $\forall xy[x \cdot (y + 1) = (x \cdot y) + x]$

and for each L-formula $\varphi(x)$, with one free variable x, the axiom

Pφ. $(\varphi(0) \wedge \forall x[\varphi(x) \to \varphi(x + 1)]) \to \forall x \varphi(x)$.

The last axiom scheme tries to capture the idea of induction. Induction says that if 0 has a certain property, and whenever n has that property, then so does $n+1$, then every natural number has that property. The axiom scheme says that for each property which can be expressed as a first-order formula.

This set of axioms, or its deductive closure, is called the theory of *Peano arithmetic*, or PA. It is easy to check each axiom individually to see that $\mathbb{N}_{\text{s-ring}} \models$ PA, and hence that PA $\subseteq$ Th($\mathbb{N}_{\text{s-ring}}$). Some of the axioms look a little strange, for example (P5) is a very limited form of the associativity axiom, and we could replace it with the full associativity axiom. In fact, the full associativity of addition can be proved from all the axioms using the induction scheme. The reason for using the limited form is that it is, at first sight, weaker, and Peano was trying to write down the weakest set of axioms from which everything else could be deduced. Can everything else about arithmetic be deduced from the axioms of PA?

Definition 6.11 A set Σ of L-sentences is said to be *complete* if, for any L-sentence σ, either $\Sigma \vdash \sigma$ or $\Sigma \vdash \neg\sigma$.

Lemma 6.12 *Let σ be any $L_{\text{s-ring}}$-sentence. Then either $\sigma \in$ Th($\mathbb{N}_{\text{s-ring}}$) or $\neg\sigma \in$ Th($\mathbb{N}_{\text{s-ring}}$). In particular,* Th($\mathbb{N}_{\text{s-ring}}$) *is complete.*

Proof If $\mathbb{N}_{\text{s-ring}} \models \sigma$, then $\sigma \in$ Th($\mathbb{N}_{\text{s-ring}}$). Otherwise, $\mathbb{N}_{\text{s-ring}} \not\models \sigma$, so $\mathbb{N}_{\text{s-ring}} \models \neg\sigma$, so $\neg\sigma \in$ Th($\mathbb{N}_{\text{s-ring}}$). Now Th($\mathbb{N}_{\text{s-ring}}$) is complete, because if $\sigma \in$ Th($\mathbb{N}_{\text{s-ring}}$), then certainly Th($\mathbb{N}_{\text{s-ring}}$) $\vdash \sigma$. □

More generally, Th($\mathcal{A}$) is complete for any L-structure $\mathcal{A}$. So the question is whether PA is complete. One of the great theorems of twentieth-century mathematics is Gödel's first incompleteness theorem, which states that PA is *not* complete. In fact, Gödel showed that whatever set of axioms about $\mathbb{N}$ we write down explicitly, it will not axiomatise all of true arithmetic.

6.4 Elementary Equivalence

Definition 6.13 Two L-structures $\mathcal{A}$ and $\mathcal{B}$ are *elementarily equivalent*, written $\mathcal{A} \equiv \mathcal{B}$, iff Th($\mathcal{A}$) = Th($\mathcal{B}$), that is, for every L-sentence σ, $\mathcal{A} \models \sigma$ iff $\mathcal{B} \models \sigma$.

So elementarily equivalent L-structures cannot be distinguished by formal sentences in the language L.

Lemma 6.14 *If $\mathcal{A}$ and $\mathcal{B}$ are isomorphic L-structures, then they are elementarily equivalent.*

Proof Let $\mathcal{A} \xrightarrow{\pi} \mathcal{B}$ be an isomorphism, and let σ be an L-sentence. Since σ has no free variables, we can apply Proposition 3.7 to σ with the empty list of variables to deduce that $\mathcal{A} \models \sigma$ if and only if $\mathcal{B} \models \sigma$. This applies for every L-sentence, so $\mathcal{A} \equiv \mathcal{B}$. □

For finite structures (structures whose domain is a finite set), isomorphism and elementary equivalence are actually the same. (See the exercises.) However, we will see later that if $\mathcal{A}$ is an infinite structure, then there are infinitely many other structures which are elementarily equivalent to $\mathcal{A}$ (and thus to each other) but which are pairwise non-isomorphic.

Exercises

6.1 Given a set Σ of L-sentences, let $\mathrm{Mod}(\Sigma)$ be the class of L-structures $\mathcal{A}$ such that $\mathcal{A} \models \Sigma$. Show that $\mathrm{Th}(\mathrm{Mod}(\Sigma))$ is the deductive closure of Σ.

6.2 Show that a theory T is complete if and only if, whenever $\mathcal{A}$ and $\mathcal{B}$ are models of T, then $\mathcal{A} \equiv \mathcal{B}$.

6.3 For each of the five structures $\mathbb{N}_{\text{s-ring}}$, $\mathbb{Z}_{\text{s-ring}}$, $\mathbb{Q}_{\text{s-ring}}$, $\mathbb{R}_{\text{s-ring}}$, $\mathbb{C}_{\text{s-ring}}$, write down a sentence which is true in that structure but not in any of the other four structures.

6.4 Show that the theory of fields is not complete.

6.5 Show that there must be a model of Peano arithmetic which is not isomorphic to the standard model, $\mathbb{N}_{\text{s-ring}}$.

6.6 Show that in any model $\mathcal{A}$ of PA, the function $+^{\mathcal{A}}$ is associative.

6.7 Let $L = \langle R, f, c \rangle$, where R is a binary relation symbol, f is a binary function symbol, and c is a constant symbol. Let $\mathcal{A}$ be an L-structure with three elements. Show that there is a single L-sentence $\sigma_{\mathcal{A}}$ such that for any L-structure $\mathcal{B}$, if $\mathcal{B} \models \sigma_{\mathcal{A}}$, then $\mathcal{B} \cong \mathcal{A}$. How would you change your proof to work for an arbitrary finite structure in an arbitrary finite language?

6.8 Suppose that $\varphi(x, y)$ is an $L_{\text{s-ring}}$-formula such that for every $n \in \mathbb{N}$, the subset $\varphi(\mathbb{N}_{\text{s-ring}}, n) = \left\{x \in \mathbb{N} \,\middle|\, \mathbb{N}_{\text{s-ring}} \models \varphi(x, n)\right\}$ of $\mathbb{N}$ is non-empty. Let $\mathcal{M} \models \mathrm{Th}(\mathbb{N}_{\text{s-ring}})$, and let $a \in \mathcal{M}$. Show that $\varphi(\mathcal{M}, a)$ has a least element.

6.9 Can you find two structures which are elementarily equivalent but not isomorphic? [This is quite difficult without some theory. Later we will see many ways to answer the question.]

7
The Complex and Real Fields

One of the main problems addressed in this book is to start with a structure $\mathcal{A}$ and to write down a complete axiomatisation of its theory. The general strategy is simple. To find a candidate set of sentences Σ which might axiomatise $\mathrm{Th}(\mathcal{A})$, start by writing down some sentences capturing properties of $\mathcal{A}$. If you can find another model $\mathcal{B} \models \Sigma$ and a sentence σ such that $\mathcal{A} \models \sigma$ and $\mathcal{B} \not\models \sigma$, then add σ to Σ to get a better candidate. When you can no longer do this, you have a plausible candidate for a complete axiomatisation.

Actually proving that a given axiomatisation is complete is, in general, a difficult problem. Gödel's First Incompleteness Theorem shows that it is impossible to give an explicit complete axiomatisation for $\mathrm{Th}(\mathbb{N}_{\text{s-ring}})$. Using this result, the same can also be shown for $\mathrm{Th}(\mathbb{Z}_{\text{ring}})$ and $\mathrm{Th}(\mathbb{Q}_{\text{ring}})$.

In this chapter we will describe complete axiomatisations for the complex and real fields. In Part III of the book we will develop one method of proving the completeness of an axiomatisation, the Łos–Vaught test. We will apply this method to prove that our axiomatisation for the complex field is complete in Chapter 30. The proof for the real field is similar but more involved and can be found in the books of Marker [Mar02, Section 3.3] and of Poizat [Poi00, Section 6.6].

The real field is a good illustration of the limitations of first-order logic. The obvious axiomatisation is in *second-order logic*, where quantification over subsets is allowed. We have to find a different axiomatisation to stay within first-order logic.

7.1 The Complex Field

To axiomatise $\mathrm{Th}(\mathbb{C}_{\text{ring}})$, we start with the axioms of fields (see Example 6.9). The field $\mathbb{C}$ has characteristic 0, which means that it satisfies the axioms

$$\underbrace{1 + 1 + \cdots + 1}_{n} \neq 0$$

for each $n \in \mathbb{N}^+$. Since there are fields with non-zero characteristic, we need these axioms.

In L_{ring} we can express statements about polynomials. The fundamental theorem of algebra states that every non-constant polynomial with coefficients from $\mathbb{C}$ has a root in $\mathbb{C}$. This fact can be written as a list of sentences in L_{ring}, taking the sentence σ_n given by

$$\forall y_0 \ldots y_{n-1} \exists x[x^n + y_{n-1} \cdot x^{n-1} + \cdots + y_1 \cdot x + y_0 = 0]$$

for each $n \in \mathbb{N}^+$, where we use the usual abbreviation of x^n for multiplying n copies of x together, and omit unnecessary brackets.

Theorem 7.1 *The axioms of fields of characteristic 0 together with the axioms σ_n above axiomatise the complete theory of the complex field $\mathbb{C}_{\text{ring}}$.*

The proof will be given in Chapter 30. These axioms have become so useful that they have a name.

Definition 7.2 A field satisfying the sentences σ_n for all $n \in \mathbb{N}^+$ is said to be *algebraically closed.*

The complete theory of $\mathbb{C}_{\text{ring}}$ is known as ACF_0, the theory of algebraically closed fields of characteristic 0. Since the theory is complete, any two algebraically closed fields of characteristic 0 satisfy exactly the same L_{ring}-sentences.

7.2 Complete Ordered Fields

We now turn to the real field.

Definition 7.3 (Ordered field) The language of ordered rings is $L_{\text{o-ring}} = L_{\text{ring}} \cup \{<\}$. In this language, the axioms of ordered fields are the axioms of fields together with axioms stating that $<$ is a linear order:

O1. (Transitivity) $\forall xyz[(x < y \wedge y < z) \to x < z]$;
O2. (Linearity) $\forall xy[x < y \vee x = y \vee y < x]$;
O3. (Irreflexivity) $\forall x[\neg x < x]$;
and axioms relating the field structure with the order:
O4. (Additivity) $\forall xyz[x < y \to x + z < y + z]$; and
O5. (Multiplicativity) $\forall xyz[(x < y \wedge 0 < z) \to x \cdot z < y \cdot z]$.

An ordered field F is necessarily of characteristic 0, since from the axioms we can deduce

$$0 < 1 < 1 + 1 < 1 + 1 + 1 < \cdots,$$

so we have $\mathbb{N}$ sitting inside F as a subset in the obvious way.

The real field satisfies a further property: the completeness of its ordering.

Definition 7.4 An ordered set F is *complete* if every non-empty subset $S \subseteq F$ which has an upper bound has a least upper bound.

This is a different usage of the word *complete* from completeness of a theory. A simple consequence of the completeness property for an ordered field is that it satisfies the property of Archimedes.

Proposition 7.5 *If F is a complete ordered field, then it is Archimedean, that is, the subset $\mathbb{N}$ of F is not bounded above.*

Proof Suppose for a contradiction that $\mathbb{N}$ is bounded above. Then, by the completeness property, there is a least upper bound b for $\mathbb{N}$. Now $b - 1 < b$ so $b - 1$ is not an upper bound for $\mathbb{N}$; hence there is a natural number $n \in \mathbb{N}$ with $b - 1 < n$. But then $n + 1 \in \mathbb{N}$ and $b < n + 1$, contradicting b being an upper bound for $\mathbb{N}$. □

A very important theorem for the foundations of analysis is that there is only one complete ordered field.

Fact 7.6 *Up to isomorphism, there is exactly one complete ordered field.*

The unique (up to isomorphism) model of the axioms of complete ordered fields is the field of real numbers. There are three common ways to build models for these axioms, that is, to construct the field of real numbers. The most familiar way is to construct $\mathbb{R}$ as the field of decimal numbers. It can also be constructed via Dedekind cuts or by Cauchy sequences. These methods give three different models of the axioms, but by the above fact, they are all isomorphic, so it does not matter which of the three (or any other model constructed by a different method) is used.

7.3 First-Order Axioms for the Real Field

The axioms for ordered fields are first-order sentences of the language $L_{\text{o-ring}}$. We can write the completeness axiom in logical symbols. The formula

$$\forall x[x \in S \rightarrow x \leqslant y]$$

states that y is an upper bound for S. Abbreviating it by $\mathrm{UB}(S,y)$, the formula

$$\mathrm{UB}(S,z) \wedge \forall y[\mathrm{UB}(S,y) \rightarrow z \leqslant y]$$

states that z is the least upper bound for S. Abbreviating that by $\mathrm{LUB}(S,z)$, the sentence

$$(\forall S \subseteq F)\,[((S \neq \emptyset) \wedge \exists y[\mathrm{UB}(S,y)]) \rightarrow \exists z[\mathrm{LUB}(S,z)]]$$

expresses the completeness axiom. However, this is *not* a formula in the language $L_{\text{o-ring}}$ as we have defined it because we have also used the symbols $\in$, $\subseteq$, and $\emptyset$, but most importantly, we have a variable S which does not range over the elements of F but over its subsets, and then we have quantified over that. This is not allowed in first-order logic, but it is allowed in so-called *second-order* logic. While we can make sense of what this axiom means, it is actually not expressible by any set of first-order sentences. (See Exercise 9.5.) It follows that there are models of the complete first-order theory of the real ordered field, $\mathrm{Th}(\mathbb{R}_{\text{o-ring}})$, which are necessarily ordered fields but which are not complete as ordered sets.

However, it turns out that we can axiomatise the first-order theory $\mathrm{Th}(\mathbb{R}_{\text{o-ring}})$ in a way closer to the axiomatisation of algebraically closed fields.

Definition 7.7 The axioms for *real-closed fields* consist of the axioms of ordered fields together with

RCF1. $\forall x[0 \leqslant x \rightarrow \exists y[y^2 = x]]$,
RCF2. $\forall y_0 \dots y_{n-1} \exists x[x^n + y_{n-1} \cdot x^{n-1} + \cdots + y_1 \cdot x + y_0 = 0]$
for each *odd* $n \in \mathbb{N}$.

In other words, a real-closed field is an ordered field in which every positive element has a square root and every odd-degree polynomial has a root.

Fact 7.8 *The axioms of real-closed fields axiomatise the complete theory* RCF *of the real ordered field* $\mathbb{R}_{\text{o-ring}}$.

Using the basic ideas of real analysis, it is easy to verify that $\mathbb{R}_{\text{o-ring}}$ is a real-closed field. However, the proof that the theory RCF is complete is beyond the scope of this book.

Remark 7.9 Often the theory RCF is defined as an L_{ring}-theory rather than an $L_{\text{o-ring}}$-theory. The order is definable from the field structure as $x < y \leftrightarrow \exists z[z \neq 0 \wedge x + z^2 = y]$, so it would be possible to translate all the axioms we have given involving $<$ into L_{ring}-sentences. However, other axiomatisations are also possible. See [Mar02, Section 3.3] and [Poi00, Section 6.6] for more details.

Exercises

7.1 Give an example to show that the axioms of ordered fields do not axiomatise a complete theory.

7.2 Verify that $\mathbb{R}_{\text{o-ring}}$ does satisfy the of axioms for real-closed fields, using standard theorems of real analysis.

7.3 Prove that in any ordered field F we have $0 < 1$ and $1 < 1 + 1$ and, more generally, that $\langle \mathbb{N}; +, 0, 1, < \rangle$ embeds in $\langle F; +, 0, 1, < \rangle$.

7.4 Let Σ be the set of axioms of ordered fields. Show that $\Sigma \vdash \forall x[x \cdot x \geqslant 0]$.

7.5 Give a linear order on $\mathbb{C}_{\text{ring}}$ satisfying axioms O1–O4. Show that there is no linear order on $\mathbb{C}_{\text{ring}}$ satisfying axioms O1–O5.

7.6 Suppose that $S \subseteq \mathbb{R}$ is non-empty and bounded above, and furthermore that it is defined by an $L_{\text{o-ring}}$-formula $\varphi(x)$. Now suppose that $\mathcal{R}$ is another real-closed field, with domain R. Show that the subset $\varphi(\mathcal{R})$ of R defined by $\varphi(x)$ has a least upper bound in R.

7.7 Is the usual definition of topological spaces given by first-order axioms or second-order axioms? What about metric spaces?

7.8 [This exercise requires some knowledge of field theory.] A complex number is *algebraic* if it is a zero of a non-trivial polynomial with integer coefficients. Show that the set of algebraic numbers forms a subfield of $\mathbb{C}_{\text{ring}}$ which is algebraically closed. Show also that the set of real algebraic numbers forms a subfield of $\mathbb{R}$ which is real-closed.

7.9 Look up the notion of Puiseux series, and find proofs that the field of Puiseux series over $\mathbb{C}$ forms an algebraically closed field and that the field of Puiseux series over $\mathbb{R}$ forms a real-closed field.

8
Compactness and New Constants

If we start with a non-empty class of structures C and form a theory $T = \mathrm{Th}(C)$, we know that T is satisfiable (that is, it has a model) because we have some models to start with. However, we can write down a set of sentences Σ without any particular model in mind and want to know that it is satisfiable. It may be very difficult to describe a model of Σ explicitly. However, the next theorem, which is the most important theorem in model theory, gives us at least a practical way to show that models do exist and that when an infinite model exists, then there are a lot of other models as well.

In the next chapter we will see that the same theorem helps us to understand the class of models of a theory, and it will be used to show that some classes of structures are not classes of models of any first-order theory. It has many other uses as well.

8.1 The Compactness Theorem

Theorem 8.1 (The Compactness Theorem) *Let Σ be a set of L-sentences. If every finite subset of Σ has a model, then Σ has a model.*

If every finite subset of Σ has a model, then we say that Σ is *finitely satisfiable*. So the compactness theorem says that finite satisfiability is the same as satisfiability.

There are at least three different approaches to the proof of the compactness theorem. In Chapter 11 we will see a way of explicitly constructing a model from the set of sentences Σ, which is due to Henkin. There is also a more algebraic proof, which builds a model of Σ from given models of its finite subsets using the technique of ultraproducts. See Section 4.1 of [CK90] or Chapter 4 of [Rot00]. These methods of proof are reasonably self-contained but not straightforward. By contrast, the original proof uses the idea of a formal deduction and so is not self-contained. For those who have seen the notion of a formal deduction, we give here a short proof of the compactness theorem.

Proof Given a set of L-sentences Σ and another L-sentence φ, we can write $\Sigma \vdash^d \varphi$ to mean that there is a formal deduction of φ from Σ. A formal deduction is a finite list of L-formulas, so it only uses finitely many of the sentences from Σ. So if $\Sigma \vdash^d \varphi$, then there is a finite subset Σ_0 of Σ such that $\Sigma_0 \vdash^d \varphi$. The Completeness Theorem of first-order logic states that $\Sigma \vdash^d \varphi$ if and only if $\Sigma \vdash \varphi$.

We write $\perp$ for some contradictory sentence, for example $\exists x[x \neq x]$, which is not true in any L-structure at all. Now $\Sigma \vdash \perp$ means that every model of Σ is a model of $\perp$, and since there are no models of $\perp$, it means there are no models of Σ, that is, Σ is not satisfiable.

Now suppose that Σ is not satisfiable. Then $\Sigma \vdash \perp$, so $\Sigma \vdash^d \perp$, hence there is a finite subset Σ_0 of Σ such that $\Sigma_0 \vdash^d \perp$. Then $\Sigma_0 \vdash \perp$, so Σ_0 is not satisfiable. Therefore, whenever every finite subset of Σ is satisfiable, Σ itself is satisfiable, as required. □

8.2 The Use of New Constant Symbols

We will use the compactness theorem to show that if $\mathcal{A}$ is any infinite L-structure, then there is another L-structure $\mathcal{B}$ which is elementarily equivalent to $\mathcal{A}$, that is, it satisfies exactly the same L-sentences, but which is not isomorphic to $\mathcal{A}$. The other tool we use is to change the language by adding new constant symbols.

Before proving the general theorem, we do a special case with a simpler proof which contains the main ideas.

Proposition 8.2 *There is a model of true arithmetic,* $\mathrm{Th}(\mathbb{N}_{\text{s-ring}})$*, which is not isomorphic to the* standard model $\mathbb{N}_{\text{s-ring}}$.

Proof Let $L^+ = L_{\text{s-ring}} \cup \{c\}$, where c is a new constant symbol. Let Σ be the set of L^+-sentences

$$\mathrm{Th}(\mathbb{N}_{\text{s-ring}}) \cup \{c \neq 0, c \neq 1, c \neq 1+1, c \neq 1+1+1, \dots\},$$

which consists of the complete theory of $\mathbb{N}_{\text{s-ring}}$ in the language $L_{\text{s-ring}}$, together with sentences using the new constant symbol c which say that c is not equal to any natural number. Now let Σ_0 be a finite subset of Σ. Then there is some number $m \in \mathbb{N}$ such that the sentence $c \neq \underbrace{1+1+\cdots+1}_{m}$ does not occur in Σ_0. We make a model of Σ_0 by expanding $\mathbb{N}_{\text{s-ring}}$ by interpreting the new constant symbol c as the number m. So Σ_0 is satisfiable. Hence, by compactness, there is a model $\mathcal{M}^+ = \langle M; +, \cdot, 0, 1, c^{\mathcal{M}^+} \rangle$ of Σ. We consider the *reduct* $\mathcal{M} = \langle M; +, \cdot, 0, 1 \rangle$, which is an $L_{\text{s-ring}}$-structure. Then $\mathcal{M} \models \text{Th}(\mathbb{N}_{\text{s-ring}})$, which is a complete $L_{\text{s-ring}}$-theory, and so $\text{Th}(\mathcal{M}) = \text{Th}(\mathbb{N}_{\text{s-ring}})$.

Suppose $\pi : \mathbb{N}_{\text{s-ring}} \to \mathcal{M}$ were an isomorphism. Let $a = c^{\mathcal{M}^+}$. Then there is $n \in \mathbb{N}$ such that $\pi(n) = a$. There is a term t which is either 0, 1, or of the form $1 + \cdots + 1$, such that $n = t^{\mathbb{N}_{\text{s-ring}}}$. Then $a = \pi(t^{\mathbb{N}_{\text{s-ring}}}) = t^{\mathcal{M}}$. But a is not equal to any such $t^{\mathcal{M}}$ because $\mathcal{M}^+ \models \Sigma$, so π is not surjective. Hence π does not exist, and so $\mathcal{M}$ is not isomorphic to $\mathbb{N}_{\text{s-ring}}$. □

We can make a similar argument with any infinite structure $\mathcal{A}$ in place of $\mathbb{N}_{\text{s-ring}}$, but we may need to use more than one new constant symbol.

Proposition 8.3 *Let $\mathcal{A}$ be any infinite L-structure. Then there is another L-structure $\mathcal{B}$ which is elementarily equivalent to $\mathcal{A}$ but not isomorphic to it.*

Proof Any isomorphism is a bijection, so isomorphic models have the same cardinality (size). So we will ensure that $\mathcal{B}$ has larger cardinality than $\mathcal{A}$ and so cannot be isomorphic to it. To do this, let I be a set of cardinality larger than $|A|$. Let L_I be the language obtained from L by adding distinct constant symbols c_i for each element of I.

Let $\Sigma = \text{Th}(\mathcal{A}) \cup \left\{ c_i \neq c_j \mid i, j \in I, i \neq j \right\}$, a set of L_I-sentences. We claim that Σ is finitely satisfiable. So let Σ_0 be a finite subset of Σ. Then only finitely many of the c_i, say, $c_{i_1}, \ldots, c_{i_n}$, are used in Σ_0. Now choose $a_1, \ldots, a_n$ from $\mathcal{A}$, all distinct. This is possible because $\mathcal{A}$ is an infinite structure. Expand $\mathcal{A}$ to an L_I-structure $\mathcal{A}^+$ by interpreting c_{i_j} as a_j for $j = 1, \ldots, n$, and every other c_i as a_1. Then

$$\mathcal{A}^+ \models \text{Th}(\mathcal{A}) \cup \left\{ c_{i_j} \neq c_{i_k} \mid 1 \leqslant j < k \leqslant n \right\},$$

so $\mathcal{A}^+ \models \Sigma_0$. Thus Σ is finitely satisfiable, and hence, by compactness, it has a model, say, $\mathcal{B}^+$, which is an L_I-structure. Let B be the domain of $\mathcal{B}^+$, and for each $i \in I$, let $a_i = c_i^{\mathcal{B}^+}$. Since $\mathcal{B}^+ \models \Sigma$, the elements a_i of B are all distinct, and so $|B| \geqslant |I| > |A|$.

Let $\mathcal{B}$ be the reduct of $\mathcal{B}^+$ to L. Then $\mathcal{B} \models \text{Th}(\mathcal{A})$, and so $\mathcal{B}$ is elementarily equivalent to $\mathcal{A}$. The domain of $\mathcal{B}$ is B, the same as that of $\mathcal{B}^+$. So $\mathcal{B}$ cannot be isomorphic to $\mathcal{A}$ because their cardinalities are different. □

The compactness theorem opens the door to study infinite structures by looking at different models of the same theory. We will explore this in Part III of the book. Now we give an indication of what the compactness theorem says about definable sets.

8.3 Applications to Definable Sets

Consider the closed intervals $[0, 1/n]$ on the real line, for $n \in \mathbb{N}^+$. The intersection of all these intervals is $\{0\}$, so in particular, the intersection is non-empty. If instead we consider the open intervals $(0, 1/n)$ for $n \in \mathbb{N}^+$, then the intersection is empty. However, if we take any finite sub-collection, say, $\{(0, 1/n_1), \ldots, (0, 1/n_r)\}$, then the intersection is the non-empty interval $(0, 1/N)$, where $N = \max\{n_1, \ldots, n_r\}$.

Proposition 8.4 *There is a model $\mathcal{R}$ of* $\mathrm{Th}(\mathbb{R}_{\text{o-ring}})$ *in which the intersection of the intervals* $\{(0, 1/n) \mid n \in \mathbb{N}^+\}$*, as interpreted in $\mathcal{R}$, is nonempty.*

Proof The interval $(0, 1/n)$ is defined by the $L_{\text{o-ring}}$-formula $\varphi_n(x)$ given by $0 < x \wedge n \cdot x < 1$, where we use n as an abbreviation for $\underbrace{1 + \cdots + 1}_{n}$. Let c be a new constant symbol, and let $\Sigma = \mathrm{Th}(\mathbb{R}_{\text{o-ring}}) \cup \{\varphi_n(c) \mid n \in \mathbb{N}\}$, a set of sentences in the language $L_{\text{o-ring}}$ expanded by c. Let Σ_0 be a finite subset of Σ, and let $N \in \mathbb{N}^+$ be larger than any n such that $\varphi_n(c) \in \Sigma_0$. Expand $\mathbb{R}_{\text{o-ring}}$ to $\mathbb{R}^+$ by interpreting c as $1/N$. Then $\mathbb{R}^+ \models \Sigma_0$. So Σ is finitely satisfiable, and hence, by compactness, it has a model, say, $\mathcal{R}^+$. Let $a = c^{\mathcal{R}^+}$. The reduct $\mathcal{R}$ of $\mathcal{R}^+$ to $L_{\text{o-ring}}$ is a model of $\mathrm{Th}(\mathbb{R}_{\text{o-ring}})$, and the element $a \in \mathcal{R}$ satisfies $\mathcal{R} \models 0 < a \wedge n \cdot a < 1$ for every $n \in \mathbb{N}$. Hence it lies in all the intervals $(0, 1/n)$, as interpreted in $\mathcal{R}$. □

Remarks 8.5 (i) The element $a \in \mathcal{R}$ in the above proof is called an *infinitesimal*, and $1/a$ is an infinite element of $\mathcal{R}$, that is, it is larger than any natural number. So the model $\mathcal{R}$ of $\mathrm{Th}(\mathbb{R}_{\text{o-ring}})$ is a non-Archimedean ordered field.

(ii) The word *compactness* comes from topology, where it is a very useful property of a topological space. The above proposition hints that definable sets behave like the closed sets of a compact topological space, at least if we are allowed to change models. In the exercises, this is explored further, and we will revisit this idea when we consider types in Part V. However, we will generally not use the language of topology.

Exercises

8.1 Is there a model $\mathcal{M}$ of $\mathrm{Th}(\mathbb{N}_{\text{s-ring}})$ with an element $m \in \mathcal{M}$ satisfying $0 < m < 1$?

8.2 Sketch a proof of Proposition 8.3. You should list all the key ideas and explain how they fit together, without giving all the details.

8.3 Where does the proof of Proposition 8.3 go wrong if $\mathcal{A}$ is a finite L-structure?

8.4 Show that there is a model $\mathcal{M}$ of $\mathrm{Th}(\mathbb{N}_<)$ with an infinite descending chain $a_1 > a_2 > a_3 > \cdots$ in $\mathcal{M}$.

8.5 Show that there is a model $\mathcal{M}$ of $\mathrm{Th}(\mathbb{Z}_<)$ such that $\mathbb{Q}_<$ embeds in $\mathcal{M}$.

8.6 A linearly ordered set $\mathcal{A} = \langle A; < \rangle$ is *well-ordered* if every non-empty subset of A has a least element. Show that if $\mathcal{A}$ is any infinite linearly ordered set, then there is $\mathcal{B} \equiv \mathcal{A}$ such that $\mathcal{B}$ is not well ordered. Show also that there is no well-ordered set $\mathcal{A}$ such that $\mathcal{A} \equiv \mathbb{Z}_<$.

8.7 Suppose that $\mathcal{A}$ is an infinite L-structure and $\varphi(\bar{x})$ is a formula such that $\varphi(\mathcal{A})$ is an infinite set. Show that there is a $\mathcal{B}$ elementarily equivalent to $\mathcal{A}$ such that $\varphi(\mathcal{B})$ is uncountable.

8.8 Suppose $\mathcal{A}$ is any L-structure and $\{S_i \mid i \in I\}$ is any collection of definable subsets of $\mathcal{A}$ such that for every finite subset $I_0 \subseteq I$, the intersection $\bigcap_{i \in I_0} S_i$ is non-empty. Show there is $\mathcal{B} \equiv \mathcal{A}$ such that, interpreted in $\mathcal{B}$, the intersection $\bigcap_{i \in I} S_i$ is non-empty.

8.9 Suppose Σ is a set of L-sentences and $\{\varphi_i(x) \mid i \in I\}$ is a set of L-formulas such that for every model $\mathcal{A} \models \Sigma$ and every element $a \in \mathcal{A}$ there is some $i \in I$ such that $\mathcal{A} \models \varphi_i(a)$. Show that there is a finite subset $I_0 \subseteq I$ such that $\Sigma \vdash \forall x \left[\bigvee_{i \in I_0} \varphi_i(x)\right]$.

8.10 The four-colour theorem states that for any map of finitely many countries drawn on the plane, with each country connected, it is possible to colour each country one of four colours such that no two adjacent countries have the same colour. Assuming this theorem, prove that the same holds for a map with infinitely many countries.

9

Axiomatisable Classes

The axiomatic method has been important in mathematics since Euclid. Euclid used axioms to capture fundamental properties of logic, geometry, or arithmetic that were supposed to be self-evident and from which other statements could be proved. In this book, we usually consider a structure such as $\mathbb{N}_{\text{s-ring}}$, $\mathbb{C}_{\text{ring}}$, or $\mathbb{R}_{\text{o-ring}}$, which we take as given (perhaps by some second-order description), and we look for axioms sufficient to capture its first-order theory. These axioms may not be self-evident; indeed, it may be a difficult task to find them and prove that they do hold.

In this chapter, however, axioms are used in a third way: to define a class of structures. The compactness theorem will be used to give general properties of classes which can be axiomatised by first-order sentences and to show that some classes are not axiomatisable or are axiomatisable but only by an infinite list of sentences.

We will use the statement of the compactness theorem from Chapter 8, but otherwise, this chapter does not depend on that one.

Definition 9.1 If Σ is a set of L-sentences, we write $\text{Mod}(\Sigma)$ for the class of all L-structures which are models of Σ. The class $\text{Mod}(\Sigma)$ is said to be *axiomatised by* Σ and is an *axiomatisable class*. It is *finitely axiomatisable* if it is axiomatised by a finite set of sentences.

For example, the class of all groups is axiomatised by the axioms G1, G2, and G3 given in Chapter 6. To show a class of structures *is* axiomatisable, it suffices to write down appropriate axioms.

9.1 Finite and Infinite Models

The class of all groups of size 3 is axiomatisable, since it is the class of all models of the group axioms together with the sentence

$$\exists x_1 x_2 x_3 [x_1 \neq x_2 \wedge x_1 \neq x_3 \wedge x_2 \neq x_3 \wedge \forall y[y = x_1 \vee y = x_2 \vee y = x_3]].$$

By a similar argument, the class of groups of any finite size is axiomatisable. The class of all infinite groups is axiomatisable, as can be seen by considering the sentences κ_n given by $\exists x_1, \ldots, x_n \left[\bigwedge_{1 \leqslant i < j \leqslant n} x_i \neq x_j\right]$ for $n \in \mathbb{N}^+$. It is the class $\mathrm{Mod}(\{G1, G2, G3\} \cup \{\kappa_n \mid n \in \mathbb{N}^+\})$. However, the class of finite groups is not axiomatisable.

Proposition 9.2 *Let C be an axiomatisable class with arbitrarily large finite models. That is, for every $n \in \mathbb{N}$, there is a model $\mathcal{A}_n \in C$ whose domain is a finite set of size at least n. Then C contains an infinite model.*

Proof Let Σ be a set of sentences such that $C = \mathrm{Mod}(\Sigma)$. Let $\Sigma' = \Sigma \cup \{\kappa_n \mid n \in \mathbb{N}^+\}$, where the sentences κ_n are as given above, and let Σ_0 be a finite subset of Σ'. Let $N \in \mathbb{N}$ be larger than any n such that $\kappa_n \in \Sigma_0$, which is possible because Σ_0 is finite. Then $\mathcal{A}_N \models \Sigma_0$. So Σ' is finitely satisfiable, hence by compactness it has a model, say, $\mathcal{B}$. Then $\mathcal{B}$ is an infinite structure, and $\mathcal{B} \models \Sigma$, so $\mathcal{B} \in C$. □

Corollary 9.3 *The class of finite groups is not axiomatisable.*

Proof There are arbitrarily large finite groups, for example the cyclic groups C_n of order n for each $n \in \mathbb{N}^+$. So any axiomatisable class containing all finite groups also contains some infinite groups. □

9.2 Torsion in Abelian Groups

Consider an abelian group G written additively in the language $L_{\text{adgp}} = \langle +, -, 0 \rangle$. For $g \in G$ and $n \in \mathbb{N}^+$ we write ng as an abbreviation for $\underbrace{g + \cdots + g}_{n}$, as usual.

Definition 9.4 We say that g is an *n-torsion* element of G if we have $ng = 0$. It is a *torsion element* if it is an n-torsion element for some n.

A group is *torsion-free* if its only torsion element is 0.

An abelian group is a *torsion group* if all of its elements are torsion elements.

For each $n \in \mathbb{N}^+$ the sentence φ_n given by $\forall x[nx = 0 \to x = 0]$ states that there are no n-torsion elements, except for 0. The set of axioms for abelian groups together with $\{\varphi_n \mid n \in \mathbb{N}^+\}$ axiomatises the class of torsion-free abelian groups.

Now we consider torsion groups. Every finite abelian group is a torsion group, but there are also infinite torsion groups, such as the direct sum of infinitely many finite groups.

Proposition 9.5 *The class of all torsion abelian groups is not axiomatisable.*

Proof Suppose it were axiomatised by Σ. Expand the language L_{adgp} to L' by adding one new constant symbol c. Let ψ_n be the L'-sentence $nc \neq 0$, and let $\Sigma' = \Sigma \cup \{\psi_n \mid n \in \mathbb{N}^+\}$. Let Σ_0 be a finite subset of Σ', and let $N \in \mathbb{N}^+$ be larger than any n such that $\psi_n \in \Sigma_0$. Let $\mathcal{A}$ be the L'-structure consisting of the cyclic group C_N with $c^{\mathcal{A}}$ being a cyclic generator. Then $\mathcal{A} \models \Sigma$, because C_N is a torsion abelian group, and also $\mathcal{A} \models \psi_n$ for each $n < N$, because $c^{\mathcal{A}}$ has order N. Hence $\mathcal{A} \models \Sigma_0$, so Σ_0 is satisfiable. By compactness, Σ' is satisfiable. Let G be a model of Σ'. Then G is a torsion abelian group, since $G \models \Sigma$, but G has an element c^G which is not an n-torsion element for any n, because $G \models \psi_n$. This contradiction shows that Σ could not exist, hence the class of all torsion abelian groups is not axiomatisable. □

9.3 Finite Axiomatisability

Recall that an axiomatisable class is *finitely axiomatisable* iff there is a finite set of sentences Σ which axiomatises the class. For example, the class of groups is finitely axiomatisable. However our choices of axioms for algebraically closed fields and for Peano arithmetic were not finite but involved infinite axiom schemes, as did our axiomatisation of torsion-free abelian groups. It is often useful to know whether a finite axiomatisation exists, or at least if it does not exist, to prove it. We start with an easy lemma.

Lemma 9.6 *If C is finitely axiomatisable, then it is axiomatised by a single sentence.*

Proof If C is axiomatised by $\{\sigma_1, \dots, \sigma_n\}$, then it is also axiomatised by $\bigwedge_{i=1}^{n} \sigma_i$. □

Now if for example the class of torsion-free abelian groups were finitely axiomatisable, how could we find a finite list of axioms? In fact we do not have to look through all sentences.

Proposition 9.7 *If $C = \text{Mod}(\Sigma)$ and C is finitely axiomatisable, then it is axiomatised by a finite subset of Σ.*

Proof Suppose not. Then using Lemma 9.6, we have a set Σ of sentences and a single sentence φ such that $C = \text{Mod}(\Sigma)$ and $C = \text{Mod}(\varphi)$, but C is not axiomatised by any finite subset of Σ. Let Σ_0 be a finite subset of Σ. Then $\text{Mod}(\Sigma_0) \supseteq \text{Mod}(\Sigma)$, but by assumption, $\text{Mod}(\Sigma_0) \neq \text{Mod}(\Sigma)$, so there is $\mathcal{A} \models \Sigma_0$ such that $\mathcal{A} \not\models \Sigma$. Thus $\mathcal{A} \notin C$, and so $\mathcal{A} \models \neg\varphi$. Thus $\Sigma_0 \cup \{\neg\varphi\}$ is satisfiable. This holds for any finite subset Σ_0 of Σ, so, by compactness, there is a model $\mathcal{B}$ of $\Sigma \cup \{\neg\varphi\}$. But then $\mathcal{B} \models \Sigma$ so $\mathcal{B} \in C$, and $\mathcal{B} \models \neg\varphi$, so $\mathcal{B} \notin C$, a contradiction. □

Corollary 9.8 *The class of all torsion-free abelian groups is not finitely axiomatisable.*

Proof Recall that the class of torsion-free abelian groups is axiomatised by $\{\text{G1}, \text{G2}, \text{G3}, \text{AG}\} \cup \{\varphi_n \mid n \in \mathbb{N}^+\}$. Suppose Σ_0 is a finite subset of these axioms, and let $N \in \mathbb{N}$ be larger than any n such that $\varphi_n \in \Sigma_0$. Let p be a prime such that $p \geqslant N$. Then the cyclic group C_p of order p is an abelian group and has no n-torsion for $n < p$, so $C_p \models \Sigma_0$. However, C_p has p-torsion, so Σ_0 does not axiomatise the class of torsion-free abelian groups. By Proposition 9.7, this class is not finitely axiomatisable. □

Our final application of the compactness theorem in this chapter is a result about axiomatising the complement of an axiomatisable class.

Proposition 9.9 *Let C be an axiomatisable class of L-structures, and let $\mathcal{D} = \{\mathcal{A}$ an L-structure $\mid \mathcal{A} \notin C\}$. Then $\mathcal{D}$ is axiomatisable if and only if C and $\mathcal{D}$ are finitely axiomatisable.*

Proof First suppose that C is axiomatised by Σ and $\mathcal{D}$ is axiomatised by Φ. Suppose for a contradiction that C is not finitely axiomatisable. Let Σ_0 be a finite subset of Σ, and let Φ_0 be a finite subset of Φ. Then $\text{Mod}(\Sigma_0) \supsetneq \text{Mod}(\Sigma)$, so there is $\mathcal{A} \models \Sigma_0$ such that $\mathcal{A} \notin C$. So $\mathcal{A} \in \mathcal{D}$, so $\mathcal{A} \models \Sigma_0 \cup \Phi$. In particular, $\Sigma_0 \cup \Phi_0$ is satisfiable. Then, by compactness, $\Sigma \cup \Phi$ is satisfiable, so there is a model $\mathcal{M} \models \Sigma \cup \Phi$, so $\mathcal{M} \in C$ and $\mathcal{M} \in \mathcal{D}$, a contradiction. So C is finitely axiomatisable.

Now suppose C is finitely axiomatisable. Then by Lemma 9.6 it is axiomatised by a single sentence, say, φ. Then $\mathcal{D}$ is axiomatised by $\neg\varphi$. □

Exercises

9.1 Let C be a non-empty class of L-structures. Show that $C \subseteq \text{Mod}(\text{Th}(C))$ and that C is axiomatisable iff equality holds.

9.2 Let Σ be a satisfiable set of L-sentences. Show that $\Sigma \subseteq \text{Th}(\text{Mod}(\Sigma))$ and that Σ is a theory iff equality holds.

9.3 Show that the class of groups of size less than 100 is axiomatisable and that the class of infinite groups is axiomatisable but not finitely axiomatisable.

9.4 Show using the compactness theorem and the method of new constants from the previous chapter that if C is an axiomatisable class with either arbitrarily large finite models or an infinite model, then it has infinite models of arbitrarily large cardinality.

9.5 Prove that the class of complete ordered fields is not axiomatisable. [Hint: use Remark 8.5 (i).]

9.6 Is the class of cyclic groups axiomatisable? Is it finitely axiomatisable?

9.7 An abelian group G is said to be n-divisible for some $n \in \mathbb{N}^+$ if it satisfies $\forall x \exists y[ny = x]$, so each element can be divided by n (not necessarily uniquely). G is *divisible* if it is n-divisible for all $n \in \mathbb{N}^+$. Show that the class of divisible abelian groups is axiomatisable but not finitely axiomatisable.

9.8 Show that the group $G = \langle \mathbb{C} \setminus \{0\}; \cdot, 1 \rangle$ is divisible but not torsion-free.

9.9 Let DTFAG be the theory of divisible, torsion-free abelian groups. Show that if V is a $\mathbb{Q}$-vector space considered as an L_{adgp}-structure, then $V \models$ DTFAG.

9.10 The *characteristic* of a field F is the least number $n \in \mathbb{N}^+$ such that $\underbrace{1 + \cdots + 1}_{n} = 0$, if such an n exists, and 0 otherwise. The characteristic of a field is always 0 or a prime number. Show that the class of fields of a fixed prime characteristic p is finitely axiomatisable and that the class of fields of characteristic 0 is axiomatisable but not finitely axiomatisable.

9.11 Show that the class of fields of positive characteristic is not axiomatisable.

9.12 Suppose that σ is an L_{ring}-sentence which is true in every field of characteristic 0. Show that there is $N \in \mathbb{N}$ such that if F is a field of characteristic p with $p > N$, then $F \models \sigma$.

9.13 A *graph* G consists of a set of vertices, and between each pair of distinct vertices, there may or may not be an edge. We can consider a graph as a structure for the language with one binary relation E. The vertices of the graph are the elements of the structure, and vertices v, w have an edge between them if $E(v, w)$ holds. A graph is then a model of the sentence

$$\forall xy[\neg E(x, x) \wedge (E(x, y) \rightarrow E(y, x))].$$

We say that a graph is *connected* if for every pair of vertices v, w, there is a path from v to w, which means there is a sequence of vertices $v = v_1, v_2, \ldots, v_n = w$ with an edge between v_i and v_{i+1} for each $i = 1, \ldots, n-1$.

Show that the class of connected graphs is not axiomatisable.

10
Cardinality Considerations

The model theory we cover will use very few set-theoretic tools, essentially only the basics of cardinal arithmetic. We give an intuitive explanation of what little we need, omitting the proofs. In particular, we avoid the use of ordinals. The underlying set theory we use is the standard one in mathematics, called Zermelo–Fraenkel with Choice (ZFC). Details of the definitions and proofs can be found in the first few pages of Kunen [Kun11] and Jech [Jec03]. For a more gentle approach, Halmos [Hal74] is also suitable.

10.1 Cardinality

Definition 10.1 Two sets have the same *cardinality* (size) if and only if there is a bijection between them. Given a set X, we write $|X|$ for the cardinality of X. A *cardinal* is a possible value of $|X|$.

We define $|X| \leqslant |Y|$ to mean there is an injective function from X to Y.

We will make use of four basic facts about injective functions, as follows:

Facts 10.2 *Let X and Y be sets.*

(i) *If there are injective functions $X \hookrightarrow Y$ and $Y \hookrightarrow X$, then there is a bijection between X and Y. (This is called the Schröder–Bernstein theorem.)*
(ii) *Given two sets X and Y, there is either an injective function $X \hookrightarrow Y$ or an injective function $Y \hookrightarrow X$.*
(iii) *Suppose $X \neq \emptyset$. Then there is an injective function $X \hookrightarrow Y$ if and only if there is a surjective function $Y \twoheadrightarrow X$.*

(iv) *For any set X, there is a set Y (e.g., the power set $\mathcal{P}X$) such that there is an injective function $X \hookrightarrow Y$ but no injective function $Y \hookrightarrow X$. (This is Cantor's theorem.)*

From facts (i) and (ii), we see that $\leqslant$ is a linear order on cardinals. Fact (iii) is often useful as an alternative way to show that $|X| \leqslant |Y|$. Fact (iv) says there is no greatest cardinal.

A set is *finite* precisely when its cardinality is some natural number $n \in \mathbb{N}$; otherwise, it is *infinite*. There is a smallest infinite cardinal, which is $|\mathbb{N}|$, written $\aleph_0$. ($\aleph$, pronounced aleph, is the first letter of the Hebrew alphabet.) A set is *countable* precisely when it is either finite or of cardinality $\aleph_0$; otherwise, it is *uncountable*. It is *countably infinite* if it is countable and infinite.

From Fact (iv), we see that the power set of the natural numbers, $\mathcal{P}\mathbb{N}$, is uncountable. It is actually in bijection with the set $\mathbb{R}$ of real numbers, so $\mathbb{R}$ is also uncountable.

10.2 Basic Cardinal Arithmetic

We can define addition and multiplication of cardinals. We define $|X| \cdot |Y|$ to be $|X \times Y|$, where $X \times Y$ is the usual cartesian product of the sets given by $\{(x, y) \mid x \in X, y \in Y\}$. We define $|X| + |Y|$ to be $|X \times \{0\} \cup Y \times \{1\}|$. Doing basic arithmetic with infinite cardinals is much easier than it is to do with natural numbers. Using Facts 10.2, one can prove the following results about cardinal arithmetic.

Fact 10.3 *If both X and Y are finite, with $|X| = n$ and $|Y| = m$, then $|X| + |Y| = n + m$ and $|X| \cdot |Y| = nm$, the usual sum and product of natural numbers. If at least one of X and Y is infinite and both are nonzero, then*

$$|X| + |Y| = |X| \cdot |Y| = \max(|X|, |Y|).$$

Fact 10.4 *Suppose that κ is an infinite cardinal and, for each $n \in \mathbb{N}$, X_n is a set such that $|X_n| \leqslant \kappa$. Let $X = \bigcup_{n \in \mathbb{N}} X_n$. Then $|X| \leqslant \kappa$.*

We give an application of this fact.

Proposition 10.5 *If X is a nonempty set and S is the set of finite strings of elements of X, that is, $S = \bigcup_{n \in \mathbb{N}} X^n$, then $|S| = \max(|X|, \aleph_0)$.*

Proof Take $x \in X$. Define $f : \mathbb{N} \to S$ by taking $f(n)$ to be the string of length n with x in every place. Then f is injective and shows that $|S| \geqslant \aleph_0$.

The inclusion map $X \hookrightarrow S$ which takes an element of X to itself as a string of length 1 shows that $|S| \geqslant |X|$. So $|S| \geqslant \max(|X|, \aleph_0)$.

For any set X, $|X^0|$ just consists of the unique string of length 0, so $|X^0| = 1$. If X is finite, then $|X^n|$ is finite for all $n \in \mathbb{N}$; otherwise, by Fact 10.3 and induction on n, we see that $|X^n| = |X|$ for all $n \in \mathbb{N}$. So, by Fact 10.4, $|S| \leqslant \max(|X|, \aleph_0)$. Since cardinals are linearly ordered, we have $|S| = \max(|X|, \aleph_0)$, as required. □

That completes our brief survey of cardinal arithmetic.

10.3 The Cardinality of a Language

We now consider an application to our formal languages. We define the cardinality $|L|$ of a language L to be the cardinality of the set Form(L) of L-formulas, where we only allow countably many variables, say, only x_n for $n \in \mathbb{N}$. Write Symb(L) for the set of relation, function, and constant symbols of L, that is, the vocabulary of L.

Proposition 10.6 *$|L|$ is equal to the maximum of $\aleph_0$ and $|\operatorname{Symb}(L)|$.*

Proof The set Form(L) contains all the formulas $(x_n = x_n)$ for $n \in \mathbb{N}$. This gives an injective function from $\mathbb{N}$ into Form(L), so $|\operatorname{Form}(L)| \geqslant \aleph_0$.

Considering the atomic formulas $c = x_0$, $f(x_1, \ldots, x_n) = x_0$, and $R(x_1, \ldots, x_n)$, we can see that for every symbol in Symb(L), there is a formula containing it and no other symbol from Symb(L). So there is an injective function from Symb(L) to Form(L), and hence $|\operatorname{Form}(L)| \geqslant |\operatorname{Symb}(L)|$. Thus $|\operatorname{Form}(L)| \geqslant \max(|\operatorname{Symb}(L)|, \aleph_0)$.

Now let $X = \operatorname{Symb}(L) \cup \{=, \wedge, \neg, \exists, (,), [,]\} \cup \{x_n \mid n \in \mathbb{N}\}$. Then

$$|X| = |\operatorname{Symb}(L)| + 8 + \aleph_0 = \max(|\operatorname{Symb}(L)|, \aleph_0).$$

Now Form(L) is a subset of the set of all finite strings of elements of X, and hence $|\operatorname{Form}(L)| \leqslant \max(|\operatorname{Symb}(L)|, \aleph_0)$ by Proposition 10.5. So $|\operatorname{Form}(L)| = \max(|\operatorname{Symb}(L)|, \aleph_0)$, as required. □

10.4 Further Set Theory

Many parts of model theory make use of ordinals and transfinite induction. A proper account of the facts of cardinal arithmetic we have stated also requires them. In this book we avoid ordinals and skirt around transfinite induction in the one or two places it is needed, except in Part VI. Readers of that part of the book should have no difficulty in learning how to do transfinite induction from another source.

Some methods in model theory require an understanding of cardinal exponentiation and issues relating to the continuum hypothesis and inaccessible cardinals. These are briefly touched on in Chapter 27.

Exercises

10.1 What is the cardinality of the set of L-terms, given the vocabulary Symb(L) of L?

10.2 Assume that for all infinite sets Y, we have $|Y \times Y| = |Y|$. Then, using Facts 10.2, deduce Facts 10.3 and 10.4.

10.3 Look up the definition of cardinal exponentiation and show that for any set X, $|\mathcal{P}X| = 2^{|X|}$.

10.4 Prove Fact 10.2(i), the Schröder–Bernstein theorem.

10.5 Suppose that $f : \mathcal{P}X \hookrightarrow X$ is an injective function. Let $A = \{f(S) \mid S \subseteq X \text{ and } f(S) \notin S\}$, and let $a = f(A)$. By considering whether $a \in A$, prove Fact 10.2(iv).

10.6 Show that Fact 10.2(iii) follows from the Axiom of Choice.

10.7 Using some form of the Axiom of Choice, for example Zorn's lemma, prove Fact 10.2 (ii).

11
Constructing Models from Syntax

The goal of this chapter is to give a proof of the compactness theorem which does not go via the notion of formal deductions.

In mathematical practice, there is a distinction between the language we use to describe mathematical objects and the objects themselves. Mathematical logic makes this distinction formal. *Syntax* is the word describing the formal language we use, as distinct from the meaning, or *semantics*, which arises when we interpret the syntax in mathematical structures. In logic it is important to be able to distinguish between the two, so for example we can distinguish between an element of a structure and a constant symbol which names the element.

However, mathematical logic also considers syntax in such a formal way that we can reason mathematically about it and indeed build mathematical structures out of it. This is precisely what we do in this chapter, where we follow Henkin's method to prove that a set Σ of sentences is satisfiable by building a model of Σ using the sentences themselves. This proof is extended in Chapter 24 to prove the omitting types theorem.

11.1 Lindenbaum's Lemma

Recall the following definitions.

Definition 11.1 A set Σ of L-sentences is

finitely satisfiable if every finite subset of Σ has a model,

deductively closed if, for every L-sentence φ, if $\Sigma \vdash \varphi$, then $\varphi \in \Sigma$, and

complete if, for every L-sentence φ, either $\Sigma \vdash \varphi$ or $\Sigma \vdash \neg\varphi$.

Note in particular that if Σ is complete and deductively closed, then for each L-sentence φ, either $\varphi \in \Sigma$ or $\neg\varphi \in \Sigma$, and if Σ is also finitely satisfiable, then it cannot contain both φ and $\neg\varphi$.

Lemma 11.2 (Lindenbaum's lemma) *If Σ is a finitely satisfiable set of L-sentences, then there is a finitely satisfiable, deductively closed and complete set $\Sigma^\sharp$ of L-sentences such that $\Sigma \subseteq \Sigma^\sharp$.*

The basic idea of the proof is to keep adding more sentences to Σ maintaining finite satisfiability until, for every sentence φ, either φ or $\neg\varphi$ is included. We leave the details as an exercise.

11.2 Henkin Theories

Recall that an L-term is said to be *closed* if it does not contain any variables. So if t is a closed L-term and $\mathcal{A}$ is an L-structure, then $t^{\mathcal{A}}$ is an element of A. Also recall that a theory is a satisfiable and deductively closed set of sentences.

Definition 11.3 A set Σ of L-sentences is called a *Henkin theory* if it is finitely satisfiable, deductively closed, and complete and it has the *witness property*, that is, for any L-sentence of the form $\exists x[\varphi(x)]$ in Σ, there is a closed L-term t such that the sentence $\varphi(t)$ is in Σ.

Proposition 11.4 *If Σ is a finitely satisfiable set of L-sentences, there is a language L^* which consists of L together with new constant symbols and a set Σ^* of L^*-sentences such that Σ^* is a Henkin theory.*

In order to build our Henkin theory Σ^*, we will expand the language to introduce new constant symbols, which will be the closed terms needed for the witness property. We will also use Lindenbaum's lemma to ensure we get a deductively closed and complete set of sentences. We have to iterate the process to get all the properties we want, so in fact we will define a tower of languages

$$L = L_0 \subseteq L_1 \subseteq L_2 \subseteq \cdots \subseteq L_n \subseteq \cdots$$

and a tower of sets of sentences

$$\Sigma \subseteq \Sigma_0 \subseteq \Sigma_1 \subseteq \Sigma_2 \subseteq \cdots \subseteq \Sigma_n \subseteq \cdots,$$

with each Σ_n being a finitely satisfiable, deductively closed and complete set of L_n-sentences. Then $L^* = \bigcup_{n\in\mathbb{N}} L_n$ and $\Sigma^* = \bigcup_{n\in\mathbb{N}} \Sigma_n$ will have the desired properties. With the explanation out of the way, we start the proof.

Proof Let $L_0 = L$, and using Lindenbaum's lemma, we can take Σ_0 to be a finitely satisfiable, deductively closed and complete set of L_0-sentences extending Σ.

Assume inductively that we have L_n and Σ_n, which is a finitely satisfiable, deductively closed and complete set of L_n-sentences. Let L_{n+1} be L_n together with a new constant symbol c_φ for each L_n-sentence of the form $\exists x[\varphi(x)]$ which is in Σ_n. Let Σ'_n be Σ_n together with all sentences of the form $\exists x[\varphi(x)] \rightarrow \varphi(c_\varphi)$ for the new constant symbols.

We claim that Σ'_n is finitely satisfiable. To prove this, let S' be a finite subset of Σ'_n, and let $S = S' \cap \Sigma_n$. By inductive hypothesis, Σ_n is finitely satisfiable, so S has a model $\mathcal{A}$ which is an L_n-structure. Expand it to an L_{n+1} structure $\mathcal{A}'$ by interpreting the constant symbol c_φ as an element a of A such that $\mathcal{A} \models \varphi(a)$ if one exists, and as any element of A otherwise. Then $\mathcal{A}' \models S'$ by construction, so Σ'_n is finitely satisfiable.

Applying Lindenbaum's lemma again, we take Σ_{n+1} to be some finitely satisfiable, deductively closed and complete set of L_{n+1}-sentences extending Σ'_n.

Now take $L^* = \bigcup_{n\in\mathbb{N}} L_n$ and $\Sigma^* = \bigcup_{n\in\mathbb{N}} \Sigma_n$. Clearly Σ^* is a set of L^*-sentences. If φ is an L^*-sentence, then for some $n \in \mathbb{N}$ it is an L_n-sentence, so either φ or $\neg\varphi$ is in Σ_n, because Σ_n is deductively closed and complete, and hence either φ or $\neg\varphi$ is in Σ^*, so Σ^* is deductively closed and complete. If $\exists x[\varphi(x)] \in \Sigma^*$, then it is in some Σ_n, and then $\exists x[\varphi(x)] \rightarrow \varphi(c_\varphi) \in \Sigma_{n+1}$, and then, since Σ_{n+1} is deductively closed, $\varphi(c_\varphi) \in \Sigma_{n+1}$. So $\varphi(c_\varphi) \in \Sigma^*$, and thus Σ^* has the witness property. Finally, if S is a finite subset of Σ^*, then it is contained in some Σ_n, and Σ_n is finitely satisfiable, so S has a model, and thus Σ^* is finitely satisfiable. So Σ^* is a Henkin theory, as required. □

11.3 Canonical Models

Definition 11.5 An L-structure $\mathcal{A}$ is said to be *canonical* if, for every element $a \in A$, there is a closed L-term t such that $t^{\mathcal{A}} = a$. A *canonical model* of a theory T is a model of T which is a canonical structure.

The structures $\mathbb{Z}_{\text{ring}}$ and $\mathbb{N}_{\text{succ}}$ are canonical structures, but $\mathbb{R}_{\text{o-ring}}$ is not. If $\mathcal{A}$ is an L-structure, then we can name every element of A with a new constant symbol to get a canonical model in the expanded language L_A. In general, a theory will not have a canonical model unless it has enough closed terms, which may involve adding new constant symbols. This is precisely what Henkin theories are for.

Proposition 11.6 *Every Henkin theory has a canonical model.*

Proof Let T be a Henkin theory in the language L, and let Λ be the set of closed L-terms. Define a relation $\sim$ on Λ by $t_1 \sim t_2$ if the sentence $t_1 = t_2$ is in T. Then $\sim$ is an equivalence relation (see Exercise 11.3). Let A be the set of equivalence classes. It will be the domain of our model. We will write $\tilde{t}$ for the equivalence class of t.

Now we have to make A into an L-structure $\mathcal{A}$ by giving interpretations of the constant symbols, function symbols, and relation symbols of L. If c is a constant symbol, we define $c^{\mathcal{A}}$ to be the equivalence class of c. If f is a function symbol of arity n and $t_1, \ldots, t_n$ are closed terms of L, we define $f^{\mathcal{A}}(\tilde{t}_1, \ldots, \tilde{t}_n)$ to be the equivalence class of the closed term $f(t_1, \ldots, t_n)$. If $s_1, \ldots, s_n$ are closed terms with $s_i \sim t_i$ for each i, since $s_i = t_i$ is in T and T is deductively closed and complete, we have $f(s_1, \ldots, s_n) = f(t_1, \ldots, t_n) \in T$. So $f^{\mathcal{A}}$ is well defined. Finally, if R is a relation symbol of L of arity n and $t_1, \ldots, t_n$ are closed terms, we define $(\tilde{t}_1, \ldots, \tilde{t}_n) \in R^{\mathcal{A}}$ if and only if $R(t_1, \ldots, t_n)$ is in T. A similar argument shows that $R^{\mathcal{A}}$ is well defined.

By construction, $\mathcal{A}$ is a canonical L-structure. It remains to show that $\mathcal{A} \models T$. We prove by induction on the construction of L-formulas $\varphi(\bar{x})$ that for any $(\tilde{t}_1, \ldots, \tilde{t}_n)$ in A^n,

$$\mathcal{A} \models \varphi(\tilde{t}_1, \ldots, \tilde{t}_n) \text{ if and only if } \varphi(t_1, \ldots, t_n) \in T.$$

For atomic formulas it holds by definition. The $\wedge$ and $\neg$ inductive steps are straightforward (see Exercise 11.4).

Suppose that $\varphi(\bar{x})$ has the form $\exists y[\theta(y, \bar{x})]$. Then

$$\begin{aligned} \mathcal{A} \models \varphi(\tilde{t}_1, \ldots, \tilde{t}_n) &\text{ iff there is } s \in \Lambda \text{ such that } \mathcal{A} \models \theta(\tilde{s}, \tilde{t}_1, \ldots, \tilde{t}_n) \\ &\text{ iff there is } s \in \Lambda \text{ such that } \theta(s, t_1, \ldots, t_n) \in T \end{aligned}$$

by the inductive hypothesis.

Now $\theta(s, t_1, \ldots, t_n) \vdash \exists y[\theta(y, t_1, \ldots, t_n)]$, so if there is some $s \in \Lambda$ such that $\theta(s, t_1, \ldots, t_n) \in T$, since T is deductively closed, we have $\exists y[\theta(y, t_1, \ldots, t_n)] \in T$. Conversely, if $\exists y[\theta(y, t_1, \ldots, t_n)] \in T$, then, since T is a Henkin theory, there is a closed term s such that $\theta(s, t_1, \ldots, t_n) \in T$. That completes the inductive step.

So $\mathcal{A} \models T$ as required. □

Putting everything together, we get a proof of the compactness theorem.

Proof of Theorem 8.1 Let Σ be a finitely satisfiable set of L-sentences. By Proposition 11.4, there is a Henkin theory Σ^* extending Σ in a language L^*

expanding L. By Proposition 11.6, Σ^* has a canonical model, say, $\mathcal{A}^*$. Then the reduct $\mathcal{A}$ of $\mathcal{A}^*$ to L is a model of Σ. □

11.4 The Strong Compactness Theorem

By considering the cardinalities of the languages involved, we will show that Henkin's method proves a stronger version of the compactness theorem.

Theorem 11.7 (Strong compactness theorem) *If Σ is a finitely satisfiable set of L-sentences, then there is a model $\mathcal{A}$ of Σ of cardinality at most $|L|$.*

Proof We will show that in the proof of Proposition 11.4, the language L^* has the same cardinality as L. Then, in the canonical model $\mathcal{A}$ given by Proposition 11.6, every element of $\mathcal{A}$ is named by an L^*-term, so $|A| \leqslant |L^*| = |L|$.

For each $n \in \mathbb{N}$, Σ_n is a set of $|L_n|$-sentences, so $|\Sigma_n| \leqslant |L_n|$. We have $|\operatorname{Symb}(L_{n+1})| \leqslant |\operatorname{Symb}(L_n)| + |\Sigma_n|$, so by Proposition 10.6, $|L_{n+1}| \leqslant |L_n| + |\Sigma_n| = |L_n|$. So, by induction on n, $|L_n| = |L_0| = |L|$ for all $n \in \mathbb{N}$. Since $L^* = \bigcup_{n\in\mathbb{N}} L_n$, by Fact 10.4, $|L^*| = |L|$, as required. □

Exercises

11.1 Let L be a countable language, and enumerate the set of L-sentences as $(\varphi_n)_{n\in\mathbb{N}}$. Let Σ be a finitely satisfiable set of L-sentences, and show that at least one of $\Sigma \cup \{\varphi_0\}$ or $\Sigma \cup \{\neg\varphi_0\}$ is finitely satisfiable. Then build on this idea to prove Lindenbaum's lemma in the case that L is countable.

11.2 [For those who know some set theory] Use transfinite induction or Zorn's lemma to prove Lindenbaum's lemma for a language of arbitrary cardinality.

11.3 Show that the relation $\sim$ on the set of closed L-terms of a Henkin theory T given by "$t_1 \sim t_2$ if the sentence $t_1 = t_2$ is in T" is an equivalence relation.

11.4 Complete the proof of Proposition 11.6 by showing that $R^{\mathcal{A}}$ is well defined and that the $\wedge$ and $\neg$ steps of the induction on formulas hold.

11.5 Summarise the proof of the compactness theorem; that is, list all of the key ideas and explain how they fit together, without giving all the details.

Part III

Changing Models

In this third part of the book, we extend the idea that a theory has different models with the Löwenheim–Skolem theorems. The Downward Löwenheim–Skolem theorem is proved using induction on formulas and the new idea of Skolem functions. The Upward Löwenheim–Skolem theorem is proved by extending the method of new constants to the method of diagrams.

The simplest possible description for the class of models of a theory is that the models are determined by a single cardinal invariant, which is the case for vector spaces which are determined by their dimension. In large enough models, this dimension is equal to the cardinality of the model. A theory like this with only one model of a certain cardinality is called categorical. The Łos–Vaught test gives the completeness of an axiomatisation for such categorical theories. So for several examples we are able to complete the first two parts of our programme: to give an axiomatisation for the theory of a structure and to classify the other models of the theory. The important back-and-forth method is introduced to achieve the first aim for dense linear orders, although it turns out to be impossible to classify all the models in that case. In the last chapter of this part, the whole programme is given as an extended exercise for the natural numbers with the successor function. This includes some investigation of the definable sets, setting the scene for Part IV.

12
Elementary Substructures

We now consider the problem of finding new models of the theory of a given structure. In this chapter we start with a structure $\mathcal{B}$ and find substructures $\mathcal{A}$ of $\mathcal{B}$ which are elementarily equivalent to it, and in fact cannot be distinguished from $\mathcal{B}$ even by formulas applied to elements of $\mathcal{A}$. The new model $\mathcal{A}$ is built using Skolem functions, and the method of proof is induction on the construction of formulas. The Tarski–Vaught test helps to organise the induction proof efficiently.

12.1 Elementary Substructures

We first recall the notion of a substructure from Chapter 5.

Definition 12.1 Let $\mathcal{A}$ and $\mathcal{B}$ be L-structures, and suppose the domain of $\mathcal{A}$ is a subset of the domain of $\mathcal{B}$. We write $\mathcal{A} \subseteq \mathcal{B}$ and say that $\mathcal{A}$ is a *substructure* of $\mathcal{B}$, and $\mathcal{B}$ is an *extension* of $\mathcal{A}$ iff the inclusion of $\mathcal{A}$ into $\mathcal{B}$ is an embedding of L-structures.

For example, if $\mathcal{B}$ is a group considered as an L_{gp}-structure, then the substructures of $\mathcal{B}$ are exactly the subgroups of $\mathcal{B}$.

Recall that if $\mathcal{A}$ is a substructure of $\mathcal{B}$, $\varphi(\bar{x})$ is an atomic (or even quantifier-free) formula, and $\bar{a} \in A$, then $\mathcal{A} \models \varphi(\bar{a})$ if and only if $\mathcal{B} \models \varphi(\bar{a})$. However, a sentence with quantifiers may have different truth values in $\mathcal{A}$ and $\mathcal{B}$.

Definition 12.2 We say that $\mathcal{A}$ is an *elementary substructure* of $\mathcal{B}$ and $\mathcal{B}$ is an *elementary extension* of $\mathcal{A}$, and write $\mathcal{A} \preccurlyeq \mathcal{B}$, if the domain of $\mathcal{A}$ is a subset of the domain of $\mathcal{B}$ and, for each formula $\varphi(\bar{x})$ and each $\bar{a} \in A^n$,

$$\mathcal{A} \models \varphi(\bar{a}) \text{ if and only if } \mathcal{B} \models \varphi(\bar{a}).$$

We say that an embedding of L-structures $\mathcal{A} \xrightarrow{\pi} \mathcal{B}$ is an *elementary embedding* iff $\pi(\mathcal{A}) \preccurlyeq \mathcal{B}$.

When $\mathcal{A} \preccurlyeq \mathcal{B}$, then $\mathcal{A}$ and $\mathcal{B}$ are very similar; indeed, they are indistinguishable from the point of view of the truth of L-formulas applied to elements of $\mathcal{A}$. The word *elementary* is used because the structures look the same in terms of their *elements*, that is, in terms of first-order logic.

Example 12.3 Consider $\mathbb{Z}_{\text{adgp}}$ and the substructure $\mathcal{E} = \langle E; +, -, 0\rangle$ consisting of the even numbers. Then $\mathcal{E} \not\preccurlyeq \mathbb{Z}_{\text{adgp}}$, because if $\varphi(x)$ is the formula $\exists y[y + y = x]$, then $\mathbb{Z}_{\text{adgp}} \models \varphi(2)$ but $\mathcal{E} \not\models \varphi(2)$. However, $\mathbb{Z}_{\text{adgp}} \cong \mathcal{E}$, because the map $\pi : \mathbb{Z} \to E$ given by $\pi(n) = 2n$ is an isomorphism.

Separating out formulas into sentences and those with free variables, we have two easy consequences of being an elementary substructure.

Lemma 12.4 *Suppose $\mathcal{A} \preccurlyeq \mathcal{B}$. Then $\mathcal{A} \equiv \mathcal{B}$, and for any L-formula $\varphi(x_1, \ldots, x_n)$, we have $\varphi(\mathcal{A}) = \varphi(\mathcal{B}) \cap A^n$.*

Proof Both conclusions are immediate from the definitions. □

12.2 The Tarski–Vaught Test

We now turn to finding elementary substructures $\mathcal{A}$ of an L-structure $\mathcal{B}$. The key idea is that if $\mathcal{B}$ says that there is an element with some property, then there must be such an element already in $\mathcal{A}$. The surprising thing is that this is all that is needed.

Lemma 12.5 (Tarski–Vaught test) *Suppose $\mathcal{A} \subseteq \mathcal{B}$ is an L-substructure and that for every L-formula $\varphi(\bar{x}, y)$ and every $\bar{a} \in A$ and $b \in B$ such that $\mathcal{B} \models \varphi(\bar{a}, b)$, there is $d \in A$ such that $\mathcal{B} \models \varphi(\bar{a}, d)$. Then $\mathcal{A} \preccurlyeq \mathcal{B}$.*

Proof Suppose $\mathcal{A} \subseteq \mathcal{B}$, satisfying the condition in the statement of the lemma. We prove by induction on formulas $\varphi(\bar{x})$ that for each $\bar{a}$ in A,

$$\mathcal{A} \models \varphi(\bar{a}) \text{ iff } \mathcal{B} \models \varphi(\bar{a}).$$

For φ an atomic formula, it is true by Lemma 3.6, because the inclusion map $\mathcal{A} \hookrightarrow \mathcal{B}$ is an embedding.

There are inductive steps for $\wedge$, $\neg$, and $\exists$. The first two are easy and left as an exercise. So suppose $\varphi(\bar{x})$ is $\exists y \psi(\bar{x}, y)$ and $\bar{a} \in A$. If $\mathcal{A} \models \exists y \psi(\bar{a}, y)$, then there is $d \in A$ such that $\mathcal{A} \models \psi(\bar{a}, d)$, and so by induction, $\mathcal{B} \models \psi(\bar{a}, d)$, hence $\mathcal{B} \models \exists y \psi(\bar{a}, y)$. Conversely, if $\mathcal{B} \models \exists y \psi(\bar{a}, y)$, then there is $b \in B$ such that

$\mathcal{B} \models \psi(\bar{a}, b)$. Then, using the condition in the statement of the lemma, there is $d \in A$ such that $\mathcal{B} \models \psi(\bar{a}, d)$. Then, by induction again, $\mathcal{A} \models \psi(\bar{a}, d)$, hence $\mathcal{A} \models \exists y \psi(\bar{a}, y)$. That completes the $\exists$ inductive step and the whole proof. □

12.3 The Downward Löwenheim–Skolem Theorem

Now we show that elementary substructures exist.

Theorem 12.6 (The Downward Löwenheim–Skolem theorem) *Let $\mathcal{B}$ be a L-structure and S a subset of $\mathcal{B}$. Then there is an elementary substructure $\mathcal{A} \preccurlyeq \mathcal{B}$ such that $S \subseteq A$ and $|A| \leqslant \max(|S|, |L|)$.*

Proof If $\mathcal{B}$ is the empty L-structure, then we can take $\mathcal{A} = \mathcal{B}$. Otherwise, for each formula $\varphi(\bar{x}, y)$, choose a function $g_\varphi : B^n \to B$ such that

$$g_\varphi(\bar{b}) = \begin{cases} \text{some } d \in B \text{ such that } \mathcal{B} \models \varphi(\bar{b}, d), \text{ if such a } d \text{ exists,} \\ \text{any } d \in B \text{ otherwise.} \end{cases}$$

Now let $S_0 = S$, and for $r \in \mathbb{N}$, define

$$S_{r+1} = S_r \cup \left\{ g_\varphi(\bar{b}) \;\middle|\; \varphi(\bar{x}, y) \text{ is an } L\text{-formula and } \bar{b} \in S_r \right\},$$

and let $A = \bigcup_{r \in \mathbb{N}} S_r$.

Note that if $\varphi(\bar{x}, y)$ is $f(\bar{x}) = y$ for a function symbol f from L, then g_φ is just the function $f^{\mathcal{B}}$. Also, if $\varphi(x, y)$ is $c = y$ for a constant symbol c, then g_φ is the constant function with value $c^{\mathcal{B}}$. So A is closed under the functions and constants in the language and hence is the domain of a substructure $\mathcal{A}$ of $\mathcal{B}$.

Next we show that $\mathcal{A} \preccurlyeq \mathcal{B}$. Suppose $\varphi(\bar{x}, y)$ is an L-formula, $\bar{a} \in A$, and $b \in B$ such that $\mathcal{B} \models \varphi(\bar{a}, b)$. Then also $\mathcal{B} \models \varphi(\bar{a}, g_\varphi(\bar{a}))$, and $g_\varphi(\bar{a}) \in A$. So, by the Tarski-Vaught test, $\mathcal{A} \preccurlyeq \mathcal{B}$.

Finally, we must consider the size of A. Using Proposition 10.5, we see that $|S_{r+1}| \leqslant \max(|S_r|, |L|)$, and so by induction, $|S_r| \leqslant \max(|S|, |L|)$ for each $r \in \mathbb{N}$. Then, using Fact 10.4, $|A| = \left|\bigcup_{r \in \mathbb{N}} S_r\right| \leqslant \max(|S|, |L|)$, as required. □

Example 12.7 Let $V \models$ ZFC, the axioms of set theory. The language is just $\{\in\}$, so there is a countable elementary substructure $V_0 \preccurlyeq V$. In particular, $V_0 \models$ ZFC. Now, in ZFC, we can prove that there is an uncountable set. Thus there is an apparent paradox (called the Skolem paradox) of a countable structure V_0 containing an uncountable set. It is a good exercise for those who know some set theory to see why this is not paradoxical at all.

Example 12.8 There is a countable elementary substructure $\mathcal{R}$ of the real ordered field $\mathbb{R}_{\text{o-ring}}$. Since $\mathbb{R}$ is uncountable, $\mathcal{R}$ is not isomorphic to $\mathbb{R}_{\text{o-ring}}$. In particular, it cannot be a complete ordered field, so it must have subsets $S \subseteq R$ which are non-empty and bounded above but which do not have a least upper bound in R. However, since $S \subseteq R \subseteq \mathbb{R}$, these sets S do have a least upper bound in $\mathbb{R}$.

12.4 Skolem Functions

The functions g_φ which we defined in the proof of the Downward Löwenheim–Skolem theorem are called *Skolem functions*. For the proof, it does not matter how we choose the values of these functions. We may have to invoke some set theory in the form of the axiom of choice, but the important thing is just that they exist. In some cases, there is no way to choose these Skolem functions to be definable. There are also structures which do have definable Skolem functions for every formula.

Example 12.9 Consider the complex field $\mathbb{C}_{\text{ring}}$ and the formula $\varphi(x, y)$ given by $x = y \cdot y$. Then a Skolem function $g_\varphi : \mathbb{C} \to \mathbb{C}$ is a square-root function, so for all $x \in \mathbb{C}$ we have $g_\varphi(x)^2 = x$. In particular, we must have $g_\varphi(-1)$ picking out one of $\pm i$, say, $+i$. But complex conjugation $z \mapsto \bar{z}$ is an automorphism of $\mathbb{C}_{\text{ring}}$, and if g_φ were definable, it would be preserved under all automorphisms, so we would have both $g_\varphi(-1) = +i$ and $g_\varphi(\overline{-1}) = \overline{+i}$, that is, $g_\varphi(-1) = -i$, a contradiction. So g_φ is not definable.

Example 12.10 $\mathbb{N}_{\text{s-ring}}$ has definable Skolem functions for every formula. To prove it, we use the fact that every non-empty subset of $\mathbb{N}$ has a smallest element. So for a formula $\varphi(\bar{x}, y)$ we can define a Skolem function by the formula $\mu_\varphi(\bar{x}, z)$ given by

$$(\varphi(\bar{x}, z) \wedge \forall y[\varphi(\bar{x}, y) \to z \leqslant y]) \vee (\neg\exists y[\varphi(\bar{x}, y)] \wedge z = 0),$$

where we have $\leqslant$ as an abbreviation, which is acceptable because $\leqslant$ is definable in $\mathbb{N}_{\text{s-ring}}$.

Exercises

12.1 Prove Lemma 12.4.

12.2 Suppose $\mathcal{A} \preccurlyeq \mathcal{B}$ and $\varphi(x)$ is an L-formula. Show that if $\varphi(\mathcal{B})$ is finite, then $\varphi(\mathcal{A}) = \varphi(\mathcal{B})$, and if $\varphi(\mathcal{B})$ is infinite, then $\varphi(\mathcal{A})$ is also infinite.

12.3 Why does Skolem's paradox seem paradoxical, and why is it not in fact contradictory?

12.4 In the real ordered field $\mathbb{R}_{\text{o-ring}}$, write down Skolem functions for the formulas $x_1 < y \wedge y < x_2$, $x_1 < y$, and $y < x_2$.

12.5 Suppose $S \subseteq \mathbb{R}$ is a finite subset which is definable in $\mathbb{R}_{\text{o-ring}}$. Show that S has a definable Skolem function. Then show that for every $s \in S$, the singleton $\{s\}$ is definable. [Hint: use induction on $|S|$.]

12.6 Suppose that $\mathcal{B}$ is a structure with Skolem functions given by function symbols in the language. Show that any substructure of $\mathcal{B}$ is an elementary substructure.

12.7 Suppose that $\mathcal{A} \subseteq \mathcal{B} \subseteq \mathcal{D}$ are L-structures such that $\mathcal{A} \preccurlyeq \mathcal{D}$ and $\mathcal{B} \preccurlyeq \mathcal{D}$. Prove that $\mathcal{A} \preccurlyeq \mathcal{B}$.

12.8 Sketch a proof of the Downward Löwenheim–Skolem theorem. You should give all the key ideas and explain how they fit together, but you do not need to give all the details.

13
Elementary Extensions

In this chapter we consider the opposite problem to the previous chapter: given a structure $\mathcal{A}$, how can we find elementary extensions of it? In Chapter 8 we used the compactness theorem in conjunction with new constant symbols to find new models of $\mathrm{Th}(\mathcal{A})$. Here we combine those ideas with the method of diagrams to find new models which are elementary extensions of $\mathcal{A}$.

13.1 The Method of Diagrams

Recall that if $\mathcal{A}$ and $\mathcal{B}$ are L-structures with domains A and B, respectively, then an *elementary embedding* $\mathcal{A} \xrightarrow{\pi} \mathcal{B}$ is an embedding such that for each L-formula $\varphi(\bar{x})$ and each $\bar{a} \in A^n$,

$$\mathcal{A} \models \varphi(\bar{a}) \text{ if and only if } \mathcal{B} \models \varphi(\pi(\bar{a})),$$

and that $\mathcal{A}$ is an *elementary substructure* of $\mathcal{B}$ and $\mathcal{B}$ is an *elementary extension* of $\mathcal{A}$ if $A \subseteq B$ and the inclusion map is an elementary embedding.

The method of diagrams turns these formulas applied to tuples from $\mathcal{A}$ into sentences by adding new constant symbols to the language. The terminology *diagram* is unfortunate, because these diagrams have nothing to do with pictures.

Given an L-structure $\mathcal{A}$ with domain A, we create a new language L_A by adding a new constant symbol c_a for every element $a \in A$. These new constant symbols must be distinct and different from any symbol in L. We then expand $\mathcal{A}$ to an L_A-structure $\mathcal{A}^+$ by interpreting the symbol c_a as the element a.

Definition 13.1 The *complete diagram* of $\mathcal{A}$ is $\mathrm{CDiag}(\mathcal{A}) = \mathrm{Th}_{L_A}(\mathcal{A}^+)$, the set of all L_A-sentences which are true in $\mathcal{A}^+$.

Proposition 13.2 *Let $\mathcal{B}$ be an L-structure. Then there is an elementary embedding $\mathcal{A} \xrightarrow{\pi} \mathcal{B}$ if and only if $\mathcal{B}$ can be expanded to a model of* CDiag($\mathcal{A}$).

Proof Suppose there is an elementary embedding $\mathcal{A} \xrightarrow{\pi} \mathcal{B}$. Expand $\mathcal{B}$ to $\mathcal{B}^+$ by defining $c_a^{\mathcal{B}^+} = \pi(a)$. Any $\sigma \in \text{CDiag}(\mathcal{A})$ is of the form $\varphi(c_{a_1}, \ldots, c_{a_n})$ for some $a_1, \ldots, a_n \in A$ such that $\mathcal{A} \models \varphi(a_1, \ldots, a_n)$. Since π is an elementary embedding, $\mathcal{B} \models \varphi(\pi(a_1), \ldots, \pi(a_n))$, and so $\mathcal{B}^+ \models \sigma$. Thus $\mathcal{B}^+ \models \text{CDiag}(\mathcal{A})$.

Conversely, if $\mathcal{B}$ can be expanded to a model $\mathcal{B}^+$ of CDiag($\mathcal{A}$), define $\mathcal{A} \xrightarrow{\pi} \mathcal{B}$ by $\pi(a) = c_a^{\mathcal{B}^+}$. A similar argument shows that π is an elementary embedding. □

Remark 13.3 Given a structure $\mathcal{A}$, the difference between an elementary extension $\mathcal{B}$ of $\mathcal{A}$ and an elementary embedding $\mathcal{A} \xrightarrow{\pi} \mathcal{B}$ is not usually very important. Given $\mathcal{A} \xrightarrow{\pi} \mathcal{B}$ with $B = \text{dom}(\mathcal{B})$, let D be a set disjoint from A such that there is a bijection $f : B \setminus \pi(A) \to D$. Let $B' = A \cup D$ and define a bijection $g : B \to B'$ by

$$g(b) = \begin{cases} \pi^{-1}(b) \text{ if } b \in \pi(A), \\ f(b) \text{ otherwise.} \end{cases}$$

We can make B' into an L-structure $\mathcal{B}'$ in exactly one way such that g is an isomorphism. Then $\mathcal{A} \preccurlyeq \mathcal{B}'$.

The complete diagram has a useful variant called the diagram.

Definition 13.4 The *diagram* Diag($\mathcal{A}$) of an L-structure $\mathcal{A}$ is the set of all atomic L_A-sentences and negations of atomic L_A-sentences which are true in $\mathcal{A}^+$.

Lemma 13.5 *There is an embedding of $\mathcal{A}$ into an L-structure $\mathcal{B}$ if and only if $\mathcal{B}$ can be expanded to a model of* Diag($\mathcal{A}$).

The proof is left as an exercise.

13.2 The Upward Löwenheim–Skolem Theorem

Now we apply the compactness theorem with two lots of new constants: those for the method of diagrams and those needed to ensure a model has large cardinality.

Theorem 13.6 (The Upward Löwenheim–Skolem theorem) *For any infinite L-structure $\mathcal{A}$ and any cardinal $\kappa \geqslant \max(|L|, |A|)$, there is an L-structure $\mathcal{B}$ of cardinality equal to κ such that $\mathcal{A} \preccurlyeq \mathcal{B}$.*

Proof Let I be a set of cardinality κ. Let $L_{A,I}$ be the language obtained from L by adding distinct constant symbols c_a for each element of A, and c_i for each element of I.

Let $\Sigma = \text{CDiag}(\mathcal{A}) \cup \left\{c_i \neq c_j \mid i, j \in I, i \neq j\right\}$, a set of $L_{A,I}$-sentences. We claim that Σ is finitely satisfiable. So let Σ_0 be a finite subset of Σ. Then only finitely many of the c_i, say, $c_{i_1}, \ldots, c_{i_n}$, are used in Σ_0. Now choose $a_1, \ldots, a_n$ from A, all distinct. Expand $\mathcal{A}$ to an $L_{A,I}$-structure $\mathcal{A}'$ by interpreting c_a as a for each $a \in A$, c_{i_j} as a_j for $j = 1, \ldots, n$, and every other c_i as a_1. Then $\mathcal{A}' \models \Sigma_0$. Thus Σ is finitely satisfiable, and hence, by compactness, it has a model, say, $\mathcal{D}'$, with domain D. Let $\mathcal{D}$ be the reduct of $\mathcal{D}'$ to L. Since $\mathcal{D}' \models \Sigma$, the elements $c_i^{\mathcal{D}'}$ of D are all distinct, and so $|D| \geqslant \kappa$.

Since $\mathcal{D}' \models \text{CDiag}(\mathcal{A})$, by Proposition 13.2, there is an elementary embedding $\mathcal{A} \xrightarrow{\pi} \mathcal{D}$. Using Remark 13.3, we can assume $\mathcal{A} \preccurlyeq \mathcal{D}$.

We are nearly done, but the elementary extension $\mathcal{D}$ we have found has cardinality at least κ, and we want an elementary extension of cardinality exactly κ. Let S be a subset of D containing A and of cardinality κ. Then, by Theorem 12.6, there is $\mathcal{B} \preccurlyeq \mathcal{D}$ of cardinality κ containing S and hence also containing A.

We must show $\mathcal{A} \preccurlyeq \mathcal{B}$. Let $\varphi(\bar{x})$ be a formula and $\bar{a} \in A$. Then

$$\mathcal{A} \models \varphi(\bar{a}) \text{ iff } \mathcal{D} \models \varphi(\bar{a}) \text{ iff } \mathcal{B} \models \varphi(\bar{a})$$

because $\mathcal{A} \preccurlyeq \mathcal{D}$ and $\mathcal{B} \preccurlyeq \mathcal{D}$. So $\mathcal{A} \preccurlyeq \mathcal{B}$. □

13.3 Non-standard Natural Numbers

A proper elementary extension $\mathcal{M}$ of $\mathbb{N}_{\text{s-ring}}$ is called a *non-standard model* of $\text{Th}(\mathbb{N}_{\text{s-ring}})$. An element of $M \setminus \mathbb{N}$ is called a *non-standard natural number*, while in this context, elements of $\mathbb{N}$ are called *standard natural numbers*. We give a few examples of properties these non-standard natural numbers can have.

Lemma 13.7 *Suppose $\mathbb{N}_{\text{s-ring}} \preccurlyeq \mathcal{M}$. Then any element $a \in M \setminus \mathbb{N}$ is greater than every standard natural number.*

Proof Suppose not, say, $a \leqslant n$. We use n as an abbreviation for the term $(1 + \cdots + 1)$, and we use $\leqslant$ as an abbreviation for the formula defining it.

Now $\mathbb{N} \models \forall x\left[x \leqslant n \rightarrow \bigvee_{i=0}^{n} x = i\right]$, so since $M \equiv \mathbb{N}$, we have $M \models \forall x\left[x \leqslant n \rightarrow \bigvee_{i=0}^{n} x = i\right]$, so $M \models \bigvee_{i=0}^{n} a = i$, and then a is a standard natural number, a contradiction. □

Proposition 13.8 *Let $\mathcal{M}$ be any proper elementary extension of $\mathbb{N}_{\text{s-ring}}$. Then there is a non-standard prime number in $\mathcal{M}$.*

Proof Note that the formula $\psi(y)$ given by

$$\forall x[\exists z[x \cdot z = y] \rightarrow (x = 1 \vee x = y)] \wedge y \neq 1$$

defines the set of prime numbers in $\mathbb{N}_{\text{s-ring}}$. We have

$$\mathbb{N} \models \forall x \exists y[\psi(y) \wedge x < y]$$

and so also $\mathcal{M} \models \forall x \exists y[\psi(y) \wedge x < y]$. Thus, given $a \in M \setminus \mathbb{N}$, there is $p \in M$ prime such that $a < p$. Then p is greater than every standard natural number, so it is non-standard. □

Proposition 13.9 *There is an elementary extension $\mathcal{M}$ of $\mathbb{N}_{\text{s-ring}}$ with a number $\beta \in \mathcal{M}$ with infinitely many distinct prime factors.*

Proof We give a compactness argument. Let $L_c = L_{\text{s-ring}} \cup \{c\}$, where c is a new constant symbol. Let $\Sigma = \text{Th}(\mathbb{N}_{\text{s-ring}}) \cup \{\exists x[n \cdot x = c] \mid n \in \mathbb{N}\}$, which states that c is divisible by every (standard) natural number. Let Σ_0 be a finite subset of Σ, and let N be the largest natural number n such that $\exists x[n \cdot x = c]$ appears in Σ_0. We make a model $\mathcal{A}$ of Σ_0 by taking $\mathbb{N}_{\text{s-ring}}$ and interpreting c as $N!$. Then $\mathcal{A} \models \Sigma_0$, so Σ_0 is satisfiable. By compactness, there is a model $\mathcal{M}^+$ of Σ. Let $\mathcal{M}$ be the reduct to the language $\langle +, \cdot, 0, 1 \rangle$. By Proposition 13.2, $\mathbb{N}_{\text{s-ring}} \preccurlyeq \mathcal{M}$, and the element $\beta = c^{\mathcal{M}^+}$ of $\mathcal{M}$ is divisible by every standard natural number, in particular every standard prime number. □

We finish our brief discussion of non-standard natural numbers with an observation about the twin prime conjecture, a famous problem in number theory.

Conjecture 13.10 (Twin prime conjecture) *There are infinitely many primes P in $\mathbb{N}$ such that $P + 2$ is also prime. (Such a pair $(P, P + 2)$ is called a pair of* twin primes.*)*

Proposition 13.11 *The twin prime conjecture is true if and only if there is some non-standard model of arithmetic $\mathcal{M}$ and at least one pair of non-standard twin primes in $\mathcal{M}$.*

Sketch proof Let $\varphi(x)$ be the formula $\psi(x) \wedge \psi(x+2)$, where ψ defines the set of primes as above. If the twin prime conjecture is true, then the subset of $\mathbb{N}$ defined by $\varphi(x)$ is infinite, and a compactness argument shows there is some $\mathcal{M}$ with a non-standard realisation of $\varphi(x)$. If the twin prime conjecture is false, then we can list all the realisations of $\varphi(x)$ as $n_1 = 3, n_2 = 5, n_3 = 11, \ldots, n_r$, for some finite r, and we have

$$\mathbb{N}_{\text{s-ring}} \models \forall x \left[\varphi(x) \to \bigvee_{i=1}^{r} x = n_i \right].$$

But then, whenever $\mathcal{M} \models \mathrm{Th}(\mathbb{N}_{\text{s-ring}})$, we have $\mathcal{M} \models \forall x[\varphi(x) \to \bigvee_{i=1}^{r} x = n_i]$, and hence there are no non-standard twin primes. □

At the time of writing (2018), the twin prime conjecture is a long-standing open problem, and we might hope that this reformulation of it would help to solve it. Unfortunately, it does not, because there is no easy way to construct explicit non-standard models of $\mathrm{Th}(\mathbb{N}_{\text{s-ring}})$ and study them directly. However, non-standard models are very useful for other theories where they can be constructed.

13.4 Non-standard Real Numbers

Now we turn to elementary extensions of the real field $\mathbb{R}_{\text{o-ring}}$.

Proposition 13.12 *There is a non-Archimedean elementary extension of* $\mathbb{R}_{\text{o-ring}}$.

Sketch proof Take a new constant symbol c and consider the set of sentences $\Sigma = \mathrm{CDiag}(\mathbb{R}_{\text{o-ring}}) \cup \{n < c \mid n \in \mathbb{N}\}$. A similar compactness argument to those we have considered already shows that Σ is satisfiable, so let $\mathcal{R}^+$ be a model, and let $\mathcal{R}$ be the reduct to the language $L_{\text{o-ring}}$. Then, as in the proof of the Upward Löwenheim-Skolem theorem, $\mathbb{R} \preccurlyeq \mathcal{R}$, and $c^{\mathcal{R}^+}$ witnesses that $\mathcal{R}$ is non-Archimedean. □

In particular, $\mathcal{R}$ is not a complete ordered field, yet every true first-order statement about $\mathbb{R}$ is also true in $\mathcal{R}$, including many theorems of analysis like the intermediate value theorem, but restricted to definable functions.

In fact, one can prove that every proper elementary extension of $\mathbb{R}$ is non-Archimedean. See Exercise 13.7(a).

Definition 13.13 If $\mathcal{R}$ is a non-Archimedean ordered field, we say an element a of $\mathcal{R}$ is *infinite* if $|a|$ is greater than every standard natural number, *finite* if it is not infinite, and *infinitesimal* if $|a| \neq 0$ but, for every $n \in \mathbb{N}$, we have $|a| < \frac{1}{n}$.

Warning: the notions of finite and infinite non-standard real numbers are different from the notions of finite and infinite as applied to sets.

Observe that a is infinite if and only if $\frac{1}{a}$ is infinitesimal. Let $\mathcal{R}$ be a proper elementary extension of $\mathbb{R}$, and let $r \in \mathbb{R}$. There is a cloud of numbers in $\mathcal{R}$ which are infinitesimally close to r, namely, all the numbers $\rho \in \mathcal{R}$ such that $|\rho - r|$ is infinitesimal or 0. We can use these *non-standard real numbers* to do non-standard analysis, for example to define the derivative of a function in the intuitive way it is sometimes done in a calculus course rather than the usual way of doing it in analysis. Exercise 13.7(c) illustrates this method for polynomial functions.

Exercises

13.1 Prove Lemma 13.5.

13.2 Show that there is a model $\mathcal{A}$ of $\mathrm{Th}(\mathbb{Z}_<)$ such that $\mathbb{Q}_<$ embeds in $\mathcal{A}$. Can the embedding be elementary?

13.3 Let T be a complete theory in a language L of cardinality λ, and suppose that T has an infinite model. Show that T has models of all cardinalities greater than or equal to λ and no finite models.

13.4 Complete the proof of Proposition 13.11 by writing down the compactness argument.

13.5 A *Mersenne prime* is a prime number of the form $2^n - 1$. Prove that there are infinitely many Mersenne primes if and only if there is a non-standard Mersenne prime. [You may assume that $f(n) = 2^n$ is a definable function in $\mathbb{N}_{\text{s-ring}}$. For a more difficult exercise, prove that it is definable.]

13.6 Show that there is a countable model $\mathcal{R}$ of $\mathrm{Th}(\mathbb{R}_{\text{o-ring}})$ which is non-Archimedean. Show that neither of $\mathcal{R}$ nor $\mathbb{R}_{\text{o-ring}}$ can be embedded in the other.

13.7 Let $\mathcal{R}$ be any proper elementary extension of the ordered field $\mathbb{R}_{\text{o-ring}}$ of real numbers.

(a) Prove that there is an infinitesimal in $\mathcal{R}$.

(b) Let I be the set consisting of zero and all infinitesimals in $\mathcal{R}$. Prove that I is closed under addition and multiplication and that if $\alpha \in I$ and $r \in \mathbb{R}$, then $\alpha \cdot r \in I$.

(c) Let $p(X) \in \mathbb{R}[X]$ be a polynomial with standard real coefficients. Show there is another polynomial $q(X) \in \mathbb{R}[X]$ such that whenever x is a standard real and α is infinitesimal, there is $\beta \in I$ such that

$$\frac{p(x+\alpha)-p(x)}{\alpha} = q(x) + \beta.$$

(d) Explain briefly how we can use the above result to define the derivative of a polynomial function. How does this method differ from the usual approach in analysis?

13.8 Sketch a proof of the Upward Löwenheim–Skolem theorem. You should give all the key ideas and explain how they fit together, but you do not need to give all the details.

13.9 An elementary chain of L-structures is a chain

$$\mathcal{A}_1 \preccurlyeq \mathcal{A}_2 \preccurlyeq \cdots \preccurlyeq \mathcal{A}_n \preccurlyeq \cdots .$$

Let $\mathcal{A} = \bigcup_{n \in \mathbb{N}^+} \mathcal{A}_n$ as in Exercise 5.9.

(a) Show that for each $n \in \mathbb{N}$ we have $\mathcal{A}_n \preccurlyeq \mathcal{A}$.

(b) Now suppose there is $\mathcal{B}$ such that each $\mathcal{A}_n \preccurlyeq \mathcal{B}$. Prove that $\mathcal{A} \preccurlyeq \mathcal{B}$.

13.10 This exercise uses the method of diagrams and compactness to prove that an axiomatisable class C is axiomatised by universal sentences if and only if it is closed under taking substructures. Suppose that C is axiomatisable and that whenever $\mathcal{B} \in C$ and $\mathcal{A}$ is a substructure of $\mathcal{B}$, $\mathcal{A} \in C$. Let $\Sigma = \mathrm{Th}(C)$, and let $\Sigma_\forall = \{\sigma \in \Sigma \mid \sigma \text{ is a universal sentence}\}$. Let $\mathcal{A} \models \Sigma_\forall$, and let $\varphi_1(\bar{c}_a), \ldots, \varphi_r(\bar{c}_a) \in \mathrm{Diag}(\mathcal{A})$.

(a) Show that $\Sigma \nvdash \forall \bar{x}\, [\neg \bigwedge_{i=1}^{r} \varphi_i(\bar{x})]$.

(b) Using this idea, show that $\Sigma \cup \mathrm{Diag}(\mathcal{A})$ is finitely satisfiable.

(c) Deduce that $C = \mathrm{Mod}(\Sigma_\forall)$.

(d) For the converse direction, prove that if C is axiomatised by a set of universal sentences, then C is closed under taking substructures.

14
Vector Spaces and Categoricity

From the Downward and Upward Löwenheim–Skolem theorems, we can see that if a first-order theory T has an infinite model, then it has at least one model of each infinite cardinality at least as large as $|L|$. We now consider theories where there is only one model in some infinite cardinality. Such theories are very interesting, not just because the theory determines the models as much as is possible, but also because, as we will see later, this property has strong consequences for the definable sets. Our prototypical example is that of vector spaces.

14.1 Vector Spaces as Structures

Let K be a field. The language $L_{K\text{-VS}}$ is $L_{\text{adgp}} = \langle +, -, 0\rangle$ together with one unary function symbol $\lambda\cdot$ for each $\lambda \in K$. (Note that we are using the two characters λ and $\cdot$ as a single symbol of our formal language.) We interpret $\lambda\cdot$ as scalar multiplication by λ.

Definition 14.1 The *theory of K-vector spaces*, $T_{K\text{-VS}}$, is the $L_{K\text{-VS}}$-theory given by the axioms for abelian groups, written additively in the language L_{adgp}, together with the following axioms which describe how scalar multiplication works:

VS1. $\forall x[0 \cdot x = 0 \wedge 1 \cdot x = x]$;

VS2$_{\lambda}$. $\forall xy[\lambda \cdot (x + y) = (\lambda \cdot x) + (\lambda \cdot y)]$;

VS3$_{\lambda,\lambda'}$. $\forall x[(\lambda \cdot x) + (\lambda' \cdot x) = \mu \cdot x]$, where $\mu = \lambda + \lambda'$ in K;

VS4$_{\lambda,\lambda'}$. $\forall x[\lambda \cdot (\lambda' \cdot x) = \nu \cdot x]$, where $\nu = \lambda\lambda'$ in K.

A *K-vector space* is a model of $T_{K\text{-VS}}$.

Note that axioms 2, 3, and 4 are not single $L_{K\text{-VS}}$-sentences but families of sentences. Axiom 2 is really a family of axioms indexed by elements of K, and axioms 3 and 4 are families of axioms indexed by pairs of elements of K. These are called *axiom schemes*. We cannot write these as single axioms in $L_{K\text{-VS}}$ because that would involve a universal quantifier which quantified over elements of K. However, that cannot exist in our language, which only allows us to quantify over elements of our structure, which is the vector space, not the field. (It is also possible to consider a language where some of the elements of a model are elements of the field and some are elements of the vector space, but that turns out to be like studying the field rather than studying the vector space.)

For any field K, there is a 0-dimensional K-vector space with only one element, 0. If K is a finite field, there are other finite K-vector spaces. Since we are mostly interested in infinite structures, it is useful to consider the theory of *infinite* K-vector spaces as well.

Definition 14.2 The theory of infinite K-vector spaces, $T^{\infty}_{K\text{-VS}}$, is the theory axiomatised by the axioms for K-vector spaces together with sentences saying that there are at least n distinct elements for each $n \in \mathbb{N}^{+}$.

14.2 Some Linear Algebra of Vector Spaces

Fix a field K. Concepts such as linear independence and bases are well known for finite-dimensional vector spaces but less well known for infinite-dimensional vector spaces, so we will review them.

Definition 14.3 Let V be a K-vector space. (We will use the same letter V for the vector space and its domain.)

- A subset S of V is *linearly independent* iff whenever $s_1, \ldots, s_n \in S$ are distinct and $\lambda_1, \ldots, \lambda_n \in K$ are such that $\sum_{i=1}^{n} \lambda_i s_i = 0$, then each $\lambda_i = 0$.
- A subset S of V is a *spanning set* for V iff every element of V can be written as a (finite) linear combination of elements of S. That is, for every $v \in V$, there are $n \in \mathbb{N}$, $s_1, \ldots, s_n \in S$ and $\lambda_1, \ldots, \lambda_n \in K$ such that $v = \sum_{i=1}^{n} \lambda_i s_i$.
- A *basis* of V is a linearly independent spanning set of V.

Note that there is no notion of an infinite linear combination here. That would require a notion of convergence of infinite series, which needs some topological or metric information we do not have in our algebraic setting.

There are three important facts which we will use.

Facts 14.4 (i) *Let B be a basis for V, and $v \in V$. Then there are a unique $n \in \mathbb{N}$, unique distinct $b_1, \ldots, b_n \in B$ and unique $\lambda_1, \ldots, \lambda_n \in K \setminus \{0\}$ such that $v = \sum_{i=1}^{n} \lambda_i b_i$. That is, every element in V can be written* uniquely *as a linear combination of elements of B.*

(ii) *Every linearly independent set can be extended to a basis. In particular, every vector space has a basis.*

(iii) *Any two bases of V have the same cardinality.*

Using (ii) and (iii), we can make the following definition.

Definition 14.5 The *dimension* $\dim V$ of a vector space V is the cardinality of a basis.

We next show that for any field K, there are K-vector spaces of every possible cardinal dimension.

Example 14.6 Let X be any set, and let $K^{\oplus X}$ be the set of functions $f : X \to K$ of finite support, that is, such that there are only finitely many $x \in X$ with $f(x) \neq 0$. We make $K^{\oplus X}$ into a K-vector space by pointwise operations:

$$(f + g)(x) := f(x) + g(x) \qquad \text{and} \qquad (\lambda \cdot f)(x) := \lambda f(x).$$

For each $x \in X$, let b_x be the function $X \to K$ given by

$$b_x(y) = \begin{cases} 1 & \text{if } y = x, \\ 0 & \text{otherwise.} \end{cases}$$

Then b_x has finite support, so $b_x \in K^{\oplus X}$. We claim that $B := \{b_x \mid x \in X\}$ is a basis for $K^{\oplus X}$. First we show that B is linearly independent. Suppose $f := \sum_{i=1}^{n} \lambda_i b_{x_i} = 0$, with the x_i distinct. Then for each $j = 1, \ldots, n$, we have $f(x_j) = \sum_{i=1}^{n} \lambda_i b_{x_i}(x_j) = \lambda_j$. But f is the zero function, so each $\lambda_j = 0$. So B is linearly independent. Now let $f \in K^{\oplus X}$, and suppose that $\{x_1, \ldots, x_n\}$ is the support of f, that is, the set of elements of X on which f takes a non-zero value. Let $\lambda_i = f(x_i)$. Then $f = \sum_{i=1}^{n} \lambda_i b_{x_i}$, so B is a spanning set.

There is an obvious bijection $X \to B$ given by $x \mapsto b_x$, so $|B| = |X|$. Thus $\dim K^{\oplus X} = |X|$.

Theorem 14.7 *Suppose that V_1 and V_2 are K-vector spaces. Then $V_1 \cong V_2$ if and only if* $\dim V_1 = \dim V_2$.

Sketch Proof If $\pi : V_1 \to V_2$ is an isomorphism and B is a basis for V_1, then it is an easy exercise to show that $\pi(B) = \{\pi(b) \mid b \in B\}$ is a basis for V_2. Then, since π is a bijection, it restricts to a bijection between B and $\pi(B)$, so

$\dim V_1 = \dim V_2$. Conversely, suppose that B_1 is a basis for V_1 and B_2 is a basis for V_2, and $\alpha : B_1 \to B_2$ is a bijection. Define $\pi : V_1 \to V_2$ by

$$\pi\left(\sum_{i=1}^{n} \lambda_i b_i\right) = \sum_{i=1}^{n} \lambda_i \alpha(b_i).$$

Then π is a well-defined function on all of V_1 because B_1 is a basis (using Fact 14.4(a)). It is injective because B_2 is linearly independent, and it is surjective because B_2 is a spanning set for V_2. It is also a linear map, that is, it preserves addition and scalar multiplication, which means it is an $L_{K\text{-VS}}$-isomorphism. So $V_1 \cong V_2$. □

14.3 The Cardinality of a Vector Space

Proposition 14.8 *If V is an infinite K-vector space, then*

$$|V| = \max(|K|, \dim V).$$

Proof Let B be a basis for V. The inclusion $B \hookrightarrow V$ is an injective map, so $|V| \geqslant |B| = \dim V$. Since V is infinite, it is not the zero-dimensional vector space, so $B \neq \emptyset$, and we can choose $b \in B$. Then there is an injective function $K \hookrightarrow V$ given by $\lambda \mapsto \lambda \cdot b$, so $|V| \geqslant |K|$. So $|V| \geqslant \max(|K|, \dim V)$.

Let $X = K \times B$, and consider the set S of all finite strings of X. We can define a function $f : S \to V$ by

$$f((\lambda_1, b_1), (\lambda_2, b_2), \ldots, (\lambda_n, b_n)) = \lambda_1 b_1 + \lambda_2 b_2 + \cdots + \lambda_n b_n.$$

This function is surjective because B is a basis, so, using Fact 10.2(iii), we have $|V| \leqslant |S|$. By Proposition 10.5, we have $|S| = \max(|X|, \aleph_0)$, and we also have $|X| = |K| \cdot |B|$. At least one of K and B is infinite (since otherwise V would be finite), and so

$$|S| = \max(|K|, |B|, \aleph_0) = \max(|K|, |B|).$$

Thus $|V| = \max(|K|, |B|)$, as required. □

Corollary 14.9 *Up to isomorphism, there is exactly one model of the theory of K-vector spaces of each infinite cardinality strictly greater than $|K|$.* □

14.4 Categoricity

We have seen that the theory of K-vector spaces has exactly one model (up to isomorphism) of each infinite cardinality κ such that $\kappa > |K|$. There is a name for such theories.

Definition 14.10 A theory T is *categorical in cardinality* κ or *κ-categorical* iff there is a model of T of cardinality κ and any two models of cardinality κ are isomorphic. $\aleph_0$-categorical is also called *countably categorical*. A structure $\mathcal{A}$ is said to be *κ-categorical* iff $\mathrm{Th}(\mathcal{A})$ is.

Recall that an L-theory T is *complete* iff for every L-sentence φ, either $T \vdash \varphi$ or $T \vdash \neg\varphi$. One useful consequence of categoricity is completeness of a theory.

Lemma 14.11 (Łos–Vaught test for completeness) *If an L-theory T is κ-categorical for some $\kappa \geqslant |L|$ and T has no finite models, then T is complete.*

Proof Let φ be an L-sentence, and let $\mathcal{M}$ be the model of T of cardinality κ. Then either $\mathcal{M} \models \varphi$ or $\mathcal{M} \models \neg\varphi$. Suppose that $\mathcal{M} \models \varphi$. Since every model of T of cardinality κ is isomorphic to $\mathcal{M}$, using Proposition 3.7, we see that there is no model of cardinality κ of $T \cup \{\neg\varphi\}$. But then, by the Löwenheim–Skolem theorems, there is no infinite model of $T \cup \{\neg\varphi\}$. We assume that T has no finite models, and hence there is no model of $T \cup \{\neg\varphi\}$. But then every model of T is a model of φ, that is, $T \vdash \varphi$. Otherwise, we have $\mathcal{M} \models \neg\varphi$, and then the same argument shows that $T \vdash \neg\varphi$. So T is complete. □

Corollary 14.12 *For any field K, the theory $T^{\infty}_{K\text{-VS}}$ is complete.* □

Corollary 14.13 *If K is an infinite field, for example $\mathbb{R}$, then all K-vector spaces of non-zero dimension are elementarily equivalent.*

Proof They are models of the same complete theory. □

It may seem very unexpected that $\mathbb{R}^1$, $\mathbb{R}^2$, and $\mathbb{R}^3$ are all elementarily equivalent as vector spaces. We think of geometry in 1, 2, and 3 dimensions as being very different. The conclusion to draw is that very little of that geometry is captured by the notion of an $\mathbb{R}$-vector space. Essentially it is because our language does not allow us to quantify over the scalars, $\mathbb{R}$.

Example 14.14 There is no $L_{\mathbb{R}\text{-VS}}$-formula $\varphi(x, y)$ which expresses that two vectors are linearly independent.

Proof If such a $\varphi(x, y)$ did exist, then we would have $\mathbb{R}^2 \models \exists xy[\varphi(x, y)]$, but $\mathbb{R}^1 \models \neg\exists xy[\varphi(x, y)]$, which is impossible, since $\mathbb{R}^1 \equiv \mathbb{R}^2$. □

Exercises

14.1 If K is a finite field and V is a K-vector space of dimension d, show that $|V| = |K|^d$.

14.2 If K is an infinite field, show that $T^{\infty}_{K\text{-VS}}$ is axiomatised by the axioms for K-vector spaces together with the single axiom $\exists x[x \neq 0]$.

14.3 If V is a one-dimensional K-vector space, show that $\mathrm{Aut}(V)$ is isomorphic to the multiplicative group $K^{\times}$ of non-zero elements of K.

14.4 If V is an n-dimensional K-vector space, show that $\mathrm{Aut}(V)$ is isomorphic to $\mathrm{GL}_n(K)$, the group of invertible $n \times n$ matrices with entries in K.

14.5 For any field K and any non-zero vector space V, show that there are exactly four definable subsets of V.

14.6 Show that if $a_1, \ldots, a_n$ and $b_1, \ldots, b_n$ are each linearly independent n-tuples from V, then there is an automorphism π of V such that $\pi(a_i) = b_i$ for each i.

14.7 Write $\mathbb{F}_3$ for the field with three elements, and let V be an infinite-dimensional $\mathbb{F}_3$-vector space. What are all the definable subsets of V and of V^2?

14.8 In the language $L_=$ with no relation, function, or constant symbols, let T be the empty theory, that is, (the deductive closure of) the empty set of $L_=$-sentences. What is a model of T? Show that T is κ-categorical for every cardinal κ.

14.9 Sketch a proof of Corollary 14.13, including the results used to prove it. You should give all the key ideas and explain how they fit together, but should not give all the details.

14.10 Suppose that $\mathcal{A}$ is a model of the theory DTFAG of divisible torsion-free abelian groups from Exercise 9.9. Show that for each $\lambda \in \mathbb{Q}$, there is an L_{adgp}-formula which defines multiplication by λ on $\mathcal{A}$ in such a way that $\mathcal{A}$ becomes a $\mathbb{Q}$-vector space. Deduce that the theory $\mathrm{DTFAG} \cup \{\exists x[x \neq 0]\}$ is complete.

15
Linear Orders

In this chapter we will look at some theories of linearly ordered sets, aiming to find complete axiomatisations using the method of categoricity. For dense linear orders such as $\mathbb{R}_<$ and $\mathbb{Q}_<$ we will achieve this using the important *back-and-forth* method.

Definition 15.1 A *linear order* on a set is a binary relation $<$ satisfying:

Transitivity $\forall xyz[(x < y \wedge y < z) \rightarrow x < z]$;
Irreflexivity $\forall x[\neg x < x]$; and
Linearity $\forall xy[x < y \vee x = y \vee y < x]$.

A set equipped with a linear order on it is a *linearly ordered set.*

We will use the abbreviations $x \leqslant y$, $y > x$, and $y \geqslant x$, as usual.

Examples 15.2 The ordered set of real numbers is written $\mathbb{R}_<$. Likewise, we have linearly ordered sets $\mathbb{N}_<$, $\mathbb{Z}_<$, $\mathbb{Q}_<$, and $I_<$, where I is the closed unit interval $I = [0, 1] \subseteq \mathbb{R}$.

Now we consider ways to distinguish these examples.

15.1 Endpoints

Both $\mathbb{N}_<$ and $I_<$ have a least element, so they are models of the sentence

$$\exists x \forall y[x \leqslant y],$$

while $\mathbb{R}_<$, $\mathbb{Q}_<$, and $\mathbb{Z}_<$ do not have a least element. Similarly, $I_<$ has a greatest element, so is a model of

$$\exists x \forall y[y \leqslant x],$$

while the other four examples are not. Least and greatest elements of a linearly ordered set are called *endpoints*, so a linearly ordered set is *without endpoints* if it satisfies the sentence

$$\forall x \exists yz[y < x \land x < z].$$

15.2 Discretely and Densely Ordered Sets

Another property of $\mathbb{Z}_<$ and $\mathbb{N}_<$ is that they are *discretely ordered*, which means that each element (except the greatest element, if it exists) has a next element greater than it, and also (except for the least element) a next element below it. This is captured by the sentences

$$\forall x[\exists w[w < x] \to \exists y[y < x \land \forall z[z < x \to z \leqslant y]]]$$

and

$$\forall x[\exists w[w > x] \to \exists y[y > x \land \forall z[z > x \to z \geqslant y]]].$$

Note that any finite non-empty linearly ordered set is discrete and has endpoints, and indeed it must be isomorphic to $\langle\{1, 2, 3, \ldots, n\}; <\rangle$ for some $n \in \mathbb{N}^+$. On the other hand, $\mathbb{R}_<$, $\mathbb{Q}_<$, and $I_<$ are *dense* linear orders, captured by the axiom

$$\forall xy[x < y \to \exists z[x < z \land z < y]].$$

Clearly a linearly ordered set cannot be both dense and discrete, although it is possible to be neither.

We have found sentences to distinguish three of our five examples from the others, and we will show that it is impossible to distinguish the last two, $\mathbb{R}_<$ and $\mathbb{Q}_<$, by a first-order sentence. Let DLO be the theory of dense linear orders without endpoints, as given by the axioms stating transitivity, irrelexivity, linearity, no endpoints, and density, together with a *non-triviality* axiom $\exists x[x = x]$ to rule out the empty linear order. We have $\mathbb{Q}_< \models$ DLO and $\mathbb{R}_< \models$ DLO.

15.3 Completeness of DLO

Theorem 15.3 *Any two countable models of DLO are isomorphic to each other.*

The method of proof we will use is called *back and forth*, and it has become very important in model theory. An early stage in the construction is shown in Figure 15.1.

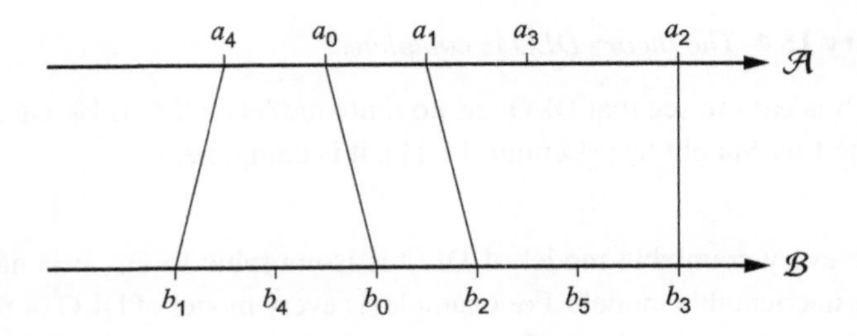

Figure 15.1 An early stage of the back-and-forth construction.

Proof of Theorem 15.3 Suppose $\mathcal{A}, \mathcal{B} \models$ DLO and both are countable. Since DLO has no finite models, $\mathcal{A}$ and $\mathcal{B}$ are both countably infinite. So we can list A as $(a_n)_{n \in \mathbb{N}}$ and B as $(b_n)_{n \in \mathbb{N}}$. We will construct an isomorphism piece by piece. In fact, for each $n \in \mathbb{N}$, we will inductively construct finite subsets $A_n \subseteq A$ and $B_n \subseteq B$ and a bijection $\pi_n : A_n \to B_n$ such that:

- for each $a, a' \in A_n$, $a < a'$ iff $\pi_n(a) < \pi_n(a')$;
- $A_n \subseteq A_{n+1}$, $B_n \subseteq B_{n+1}$, and $\pi_n \subseteq \pi_{n+1}$; and
- $a_n \in A_{2n+1}$ and $b_n \in B_{2n+2}$.

In particular, π_n is an isomorphism between $\langle A_n; < \rangle$ and $\langle B_n; < \rangle$.

We start with $A_0 = B_0 = \emptyset$, and π_0 the empty function. Now suppose n is even, say, $n = 2m$, and we have constructed A_n, B_n, and π_n. If $a_m \in A_n$, then set $A_{n+1} = A_n$, $B_{n+1} = B_n$, and $\pi_{n+1} = \pi_n$. By induction, the above conditions are satisfied. Otherwise, $a_m \notin A_n$, and we define $A_{n+1} = A_n \cup \{a_m\}$. We must find a suitable element in B to be $\pi_{n+1}(a_m)$. If n = 0, we choose $\pi_1(a_0) = b_0$. Otherwise, there are three cases. Either (i) a_m is less then every element of A_n or (ii) a_m is greater than every element of A_n or (iii) there are $c, d \in A_n$ with $c < a_m$ and $a_m < d$, and no elements of A_n strictly between c and d. In each case, because $\mathcal{B} \models$ DLO, there is $b \in B$ such that b is (i) less than every element of B_n or (ii) greater than every element of B_n or (iii) between $\pi_n(c)$ and $\pi_n(d)$. To be precise, we choose the b_r for the smallest possible value of $r \in \mathbb{N}$ with the desired property. Then we define $B_{n+1} = B_n \cup \{b_r\}$ and $\pi_{n+1} = \pi_n \cup \{(a_m, b_r)\}$.

Now suppose that n is odd, say, $n = 2m + 1$. If $b_m \in B_n$, then set $A_{n+1} = A_n$, $B_{n+1} = B_n$, and $\pi_{n+1} = \pi_n$. Otherwise, let $B_{n+1} = B_n \cup \{b_m\}$. Then we use the same process as above to find $a_r \in A$.

Now let $\pi = \bigcup_{n \in \mathbb{N}} \pi_n$. Our construction ensures that π is defined on all of A and its image is all of B. It also preserves $<$ (in particular, it is injective), and so it is an isomorphism from $\mathcal{A}$ to $\mathcal{B}$. □

Corollary 15.4 *The theory DLO is complete.*

Proof It is easy to see that DLO has no finite models. DLO is $\aleph_0$-categorical, so, by the Łos-Vaught test (Lemma 14.11), it is complete.

□

Since every countable model of DLO is isomorphic to $\mathbb{Q}_<$, it is natural to consider uncountable models. For example, is every model of DLO of the same cardinality as $\mathbb{R}$ isomorphic to $\mathbb{R}_<$? Consider a linear order $\mathcal{M}$ which consists of a copy of $\mathbb{Q}$ and a copy of $\mathbb{R}$, with every element of $\mathbb{Q}$ less than every element of $\mathbb{R}$. Then $|M| = |\mathbb{Q}| + |\mathbb{R}| = |\mathbb{R}|$.

It is easy to check that $\mathcal{M} \models$ DLO, but we can show that $\mathcal{M} \not\cong \mathbb{R}$. Indeed, suppose $\pi : \mathcal{M} \to \mathbb{R}$ were an isomorphism, and let $r = \pi(0_{\mathbb{Q}})$. Then π maps the set $\left\{a \in M \,\middle|\, a < 0_{\mathbb{Q}}\right\}$ bijectively to the set $\{b \in \mathbb{R} \mid b < r\}$, but the first set is countable and the second set is uncountable, so there cannot be a bijection between them. This contradiction shows that $\mathcal{M} \not\cong \mathbb{R}$. So DLO is not categorical in $|\mathbb{R}|$, and in fact DLO is not categorical in any uncountable cardinal.

Exercises

15.1 Give an example of a linearly ordered set which is neither discrete nor dense.

15.2 Suppose that $\mathcal{A}$ is a discrete linear order without endpoints. Show that the successor function which takes an element to the one immediately above it is definable.

15.3 Let DLO_{ep} be the theory of dense linear orders with least and greatest endpoints together with the axiom $\exists xy[x < y]$. Show that DLO_{ep} is $\aleph_0$-categorical.

15.4 Let $a, b \in \mathbb{Q}$. Show that there is an automorphism of $\mathbb{Q}_<$ taking a to b.

15.5 Let $\mathcal{A}$ be a countable linearly ordered set. Show that there is an embedding of $\mathcal{A}$ into $\mathbb{Q}_<$. [Hint: use the idea in the back-and-forth proof, but only going forth, not back.]

15.6 Show that every non-empty finite linearly ordered set is discrete and has endpoints. [Hint: use induction on the cardinality of the set.]

15.7 Show that the theory of non-empty discrete linear orders without endpoints is not $\aleph_0$-categorical.

15.8 If $\mathcal{A}$ and $\mathcal{B}$ are two linear orders, their *lexicographic product* $\mathcal{A} \times_{\text{lex}} \mathcal{B}$ is the set $A \times B$, linearly ordered by setting $(a, b) < (c, d)$ if and only if either $a < c$ or ($a = c$ and $b < d$). Show that any discrete linear order without endpoints can be written as $\mathcal{A} \times_{\text{lex}} \mathbb{Z}_<$, for some linear order $\mathcal{A}$.

15.9 Show that each singleton $\{n\}$ is definable in $\mathbb{N}_<$. Using this fact and a compactness argument, show that $\text{Th}(\mathbb{N}_<)$ is not $\aleph_0$-categorical.

15.10 Show that $\text{Th}\langle\mathbb{Z}; <, 0\rangle$ is not $\aleph_0$-categorical. Hence or otherwise show that $\text{Th}(\mathbb{Z}_<)$ is also not $\aleph_0$-categorical.

16

The Successor Structure

This chapter is a guided extended exercise devoted to studying the structure $\mathbb{N}_{\text{succ}} = \langle \mathbb{N}; s, 0 \rangle$, the natural numbers with a constant symbol naming 0, and the successor function s, a unary function defined by $s(n) = n + 1$. We will write down axioms for its theory, classify all the models of our axioms, and use the Łos–Vaught test to prove that the list of axioms is complete. We will then show how to use the other models of T_S to give information about the definable sets in $\mathbb{N}_{\text{succ}}$.

16.1 Axioms

We first observe some basic properties of the successor function:

S1. s is injective;
S2. 0 is not in the image of s; and
S3. every element of $\mathbb{N}$ other than 0 is in the image of s.

Exercise 16.1 Write down first-order sentences to capture these properties.

Slightly less obvious is that there are no 'cycles', captured by the following axiom scheme:

S4$_n$. $\forall x[s^n(x) \neq x]$ for $n \in \mathbb{N}^+$,

where $s^0(x) = x$ and $s^{n+1}(x) = s(s^n(x))$ for each $n \in \mathbb{N}$.

Let us write T_S for the set of axioms $\{\text{S1}, \text{S2}, \text{S3}\} \cup \{\text{S4}_n \mid n \in \mathbb{N}^+\}$.

Exercise 16.2 Write down a similar list of axioms that is satisfied by $\mathbb{Z}_{\text{succ}} = \langle \mathbb{Z}; s, 0 \rangle$.

Exercise 16.3 Adapt the lists of axioms to the structures $\langle \mathbb{N}; s \rangle$ and $\langle \mathbb{Z}; s \rangle$ which do not have a constant symbol naming 0.

16.2 Models

By design, $\mathbb{N}_{succ}$ is a model of T_S. By the Upward Löwenheim–Skolem theorem, there are models of T_S of every infinite cardinality. Now we will investigate what these must look like.

Exercise 16.4 Show that every model of T_S has a substructure which is isomorphic to $\mathbb{N}_{succ}$. We will identify this substructure with $\mathbb{N}$ for notational convenience.

Now suppose that $\mathcal{M} \models T_S$ and $\mathcal{M} \not\cong \mathbb{N}_{succ}$. Then there is an element $a \in M \setminus \mathbb{N}$.

Exercise 16.5 Show that a lies in a substructure of $\langle M; s \rangle$ isomorphic to $\langle \mathbb{Z}; s \rangle$.

(We have taken a reduct to forget the constant 0; otherwise, it would not be a substructure.) So any model $\mathcal{M} \models T_S$ consists of $\mathbb{N}_{succ}$ and a set of copies of $\langle \mathbb{Z}; s \rangle$. We can be more precise.

For any set I, define $\mathcal{M}_I = \langle M_I; s, 0 \rangle$ to be the structure with domain $M_I = \mathbb{N} \cup (I \times \mathbb{Z})$, the constant symbol 0 naming the zero from $\mathbb{N}$, and with s defined by $s(n) = n + 1$ for $n \in \mathbb{N}$ and $s((i, n)) = (i, n + 1)$ for each $(i, n) \in I \times \mathbb{Z}$.

Exercise 16.6 Verify that for any set I, $\mathcal{M}_I \models T_S$.

Exercise 16.7 Show that if $\mathcal{M}$ is any model of T_S, then there is a set I such that $\mathcal{M}$ is isomorphic to $\mathcal{M}_I$.

Exercise 16.8 Show that $\mathcal{M}_{I_1} \cong \mathcal{M}_{I_2}$ if and only if I_1 and I_2 have the same cardinality.

So now we have a classification of all the models of T_S, up to isomorphism.

Exercise 16.9 Show that the cardinality of $\mathcal{M}_I$ is $|I| + \aleph_0$.

Exercise 16.10 How many models are there of each cardinality, up to isomorphism? In which cardinalities is T_S categorical?

Exercise 16.11 Prove that T_S is a complete theory.

Exercise 16.12 What changes if we take $\mathbb{Z}_{\text{succ}}$, $\langle \mathbb{N}; s \rangle$, or $\langle \mathbb{Z}, s \rangle$ instead of $\mathbb{N}_{\text{succ}}$?

16.3 Definable Sets

Now we want to understand some of the definable sets in the original structure $\mathbb{N}_{\text{succ}}$. It is possible to use the method of quantifier elimination to get a complete understanding of the definable sets. See Exercise 18.4. Here we take a more direct approach.

First observe that each element $n \in \mathbb{N}$ is named by the closed term $s^n(0)$. So the singletons $\{n\}$ are all definable.

Exercise 16.13 Show that every finite subset of $\mathbb{N}_{\text{succ}}$ is definable and that every cofinite subset (that is, the complement of a finite set) is also definable.

The technique we know to show that a subset is not definable is to show that it is not preserved under some automorphism. However, $\mathbb{N}_{\text{succ}}$ itself has no automorphisms (except the identity), so we cannot use that technique directly. However, we can use it together with our understanding of the other models of T_S to deduce that certain subsets of $\mathbb{N}$ itself are not definable.

Exercise 16.14 Suppose $I \neq \emptyset$, and let $a, b \in M_I \setminus \mathbb{N}$. Show there is an automorphism π of $\mathcal{M}_I$ such that $\pi(a) = b$.

Exercise 16.15 Deduce that if S is a definable subset of M_I, then we must have either $M_I \setminus \mathbb{N} \subseteq S$ or $(M_I \setminus \mathbb{N}) \cap S = \emptyset$.

Exercise 16.16 Let $\mathcal{A}$ be any infinite structure and $\varphi(\mathcal{A}) \subseteq A$ and $\theta(\mathcal{A}) \subseteq A$ be infinite definable sets. Use a compactness argument to show there is $\mathcal{B}$, elementarily equivalent to $\mathcal{A}$, such that the subsets $\varphi(\mathcal{B})$ and $\theta(\mathcal{B})$ of B are uncountable.

Exercise 16.17 Now suppose that $\varphi(x)$ is a formula such that both $\varphi(\mathbb{N}_{\text{succ}})$ and $\neg\varphi(\mathbb{N}_{\text{succ}})$ are infinite. By considering $\varphi(\mathcal{M}_I)$ and $\neg\varphi(\mathcal{M}_I)$ for an uncountable set I, find a contradiction. Deduce that the subsets of $\mathbb{N}$ which are definable in $\mathbb{N}_{\text{succ}}$ are exactly the finite and cofinite subsets. (S is cofinite if its complement $\mathbb{N} \setminus S$ is finite.)

Similar arguments using automorphisms of the models $\mathcal{M}_I$ can be used to understand what the definable subsets of $\mathbb{N}^n$ look like for $n > 1$. They are not just finite or cofinite, for example $\{(a, a+3) \mid a \in \mathbb{N}\} \subseteq \mathbb{N}^2$ is defined by the formula $\varphi(x_1, x_2)$ given by $x_2 = s(s(s(x_1)))$. We restrict ourselves to showing that $\mathbb{N}_{\text{succ}}$ is genuinely different from $\mathbb{N}_<$.

Exercise 16.18 Take $I = \{i, j\}$, and show there is an automorphism π of $\mathcal{M}_I$ such that $\pi((i, 0)) = (j, 0)$ and $\pi((j, 0)) = (i, 0)$.

Exercise 16.19 Show that no formula can define a linear order on $\mathcal{M}_I$, and deduce that the order $<$ on $\mathbb{N}$ is not a definable subset of $\mathbb{N}^2$.

Part IV

Characterising Definable Sets

In this part of the book we reach the main theme: the study of the definable sets of a structure. We first introduce the important method of quantifier elimination. For those structures where it holds, it means that to understand the definable sets, it is enough to consider the sets defined by formulas without quantifiers. The basic back-and-forth method suffices to prove quantifier elimination for dense linear orders, and we introduce the method of substructure completeness, which works for many other examples, including vector spaces and (in Part VI) algebraically closed fields.

We then show how definable sets form Boolean algebras and relate the atoms of these Boolean algebras to principal formulas. Finally, we work out several examples, including the important case of the real field, where the definable sets are also known as semi-algebraic sets.

17

Quantifier Elimination for Dense Linear Orders

In Chapter 15 we found an axiomatisation for the complete theory of $\mathbb{Q}_<$ by using the back-and-forth method to prove countable categoricity of the theory DLO. In this chapter we will see that a minor variation of the same back-and-forth method allows us to understand what all the subsets of $\mathbb{Q}^n$ which are definable in $\mathbb{Q}_<$ are.

17.1 Principal DLO Formulas

Our first task is to understand which subsets of $\mathbb{Q}^n$ are definable with *quantifier-free* formulas $\varphi(\bar{x})$. Typically, in any structure with a reasonably chosen language, these are relatively easy, and it is quantifiers which cause the difficulties. Quantifier-free formulas are Boolean combinations of atomic formulas, which for the language $L_<$ are of the form $(x_i < x_j)$ and $(x_i = x_j)$. Examples for $n = 5$ include

$$x_1 < x_2 \wedge x_2 < x_3 \wedge x_3 = x_4 \wedge x_4 < x_5$$

and

$$x_3 < x_1 \wedge x_2 = x_4 \wedge (x_3 < x_5 \vee x_4 = x_5).$$

The first of these completely specifies the order of the variables $x_1, \ldots, x_5$, while the second only partially specifies the order. For example, it does not specify whether $x_2 < x_3$, $x_2 = x_3$, or $x_2 > x_3$. We capture the idea of completely specifying the order of the variables in the following definition.

Definition 17.1 A *principal DLO formula* for the variables $x_1, \ldots, x_n$ is a formula $\psi(x_1, \ldots, x_n)$ of the form $x_1 = x_1$ if $n = 1$ or

$$\bigwedge_{i=1}^{n-1} \left(x_{\sigma(i)} \mathbin{\square_i} x_{\sigma(i+1)}\right),$$

where $n > 1$, σ is a permutation of $\{1, \ldots, n\}$ and each $\square_i$ is either $<$ or $=$.

Evidently these formulas are quantifier-free $L_<$-formulas.

Lemma 17.2 *For each $n \in \mathbb{N}^+$, there are only finitely many principal DLO formulas for the variables $x_1, \ldots, x_n$. Furthermore, every n-tuple from $\mathbb{Q}_<$ satisfies a principal DLO formula.*

Proof The reader should attempt this as an exercise. □

17.2 Automorphisms of $\mathbb{Q}_<$

The back-and-forth method was used to produce isomophisms between any two countable dense linear orders (without endpoints). Now we use it to produce automorphisms.

Proposition 17.3 *Suppose that $\psi(\bar{x})$ is a principal DLO formula and $\bar{a}, \bar{b} \in \mathbb{Q}^n$ are such that $\mathbb{Q}_< \models \psi(\bar{a}) \wedge \psi(\bar{b})$. Then there is an automorphism π of $\mathbb{Q}_<$ such that $\pi(\bar{a}) = \bar{b}$.*

Proof We have $\bar{a} = (a_1, \ldots, a_n)$ and $\bar{b} = (b_1, \ldots, b_n)$. Extend both of these tuples to enumerations $(a_m)_{m \in \mathbb{N}^+}$ and $(b_m)_{m \in \mathbb{N}^+}$ of $\mathbb{Q}$. Let $A_n = \{a_1, \ldots, a_n\}$, let $B_n = \{b_1, \ldots, b_n\}$, and let $\pi_n : A_n \to B_n$ be given by $\pi_n(a_i) = b_i$ for $i = 1, \ldots, n$. Since $\psi(\bar{x})$ is a principal DLO formula, we have for each $1 \leqslant i, j \leqslant n$ that $a_i < a_j$ iff $\pi_n(a_i) < \pi_n(a_j)$. Now for each $m \in \mathbb{N}$ such that $m > n$ we construct π_m by the back-and-forth method, exactly as in the proof of Theorem 15.3. Then setting $\pi = \bigcup_{m \geqslant n} \pi_m$, we see that π is an automorphism of $\mathbb{Q}_<$ such that $\pi(\bar{a}) = \bar{b}$. □

17.3 Quantifier Elimination for $\mathbb{Q}_<$

Theorem 17.4 *For any $n \in \mathbb{N}^+$, any non-empty subset of $\mathbb{Q}^n$ which is definable in $\mathbb{Q}_<$ is defined by a finite disjunction of principal DLO formulas.*

Proof Let $n \in \mathbb{N}^+$, and suppose $\varphi(x_1, \ldots, x_n)$ is an $L_<$-formula such that $\varphi(\mathbb{Q}_<) \neq \emptyset$. Let P be the set of all principal DLO formulas ψ such that $\mathbb{Q}_< \models \exists\bar{x}[\varphi(\bar{x}) \wedge \psi(\bar{x})]$, and let $\Psi(\bar{x}) = \bigvee_{\psi \in P} \psi(\bar{x})$. By Lemma 17.2, the set P is finite, so $\Psi(\bar{x})$ is a first-order $L_<$-formula.

We will show that $\Psi(\bar{x})$ defines the same subset of $\mathbb{Q}^n$ as $\varphi(\bar{x})$. Suppose $\mathbb{Q}_< \models \varphi(\bar{a})$. Then $\bar{a}$ satisfies some principal DLO formula $\psi(\bar{x})$, and then $\psi \in P$ by definition of P. Since $\mathbb{Q}_< \models \psi(\bar{a})$, we have $\mathbb{Q}_< \models \Psi(\bar{a})$.

Conversely, suppose that $\mathbb{Q}_< \models \Psi(\bar{a})$. Then there is $\psi \in P$ such that $\mathbb{Q}_< \models \psi(\bar{a})$, and since $\psi \in P$, there is $\bar{b} \in \mathbb{Q}^n$ such that $\mathbb{Q}_< \models \varphi(\bar{b}) \wedge \psi(\bar{b})$. We have $\mathbb{Q}_< \models \psi(\bar{a}) \wedge \psi(\bar{b})$, so by Proposition 17.3, there is an automorphism π of $\mathbb{Q}_<$ such that $\pi(\bar{a}) = \bar{b}$. Since $\mathbb{Q}_< \models \varphi(\bar{b})$, by Proposition 3.7, we have $\mathbb{Q}_< \models \varphi(\bar{a})$.

So $\Psi(\bar{x})$ and $\varphi(\bar{x})$ define the same subset of $\mathbb{Q}^n$, as required. □

Definition 17.5 A structure $\mathcal{A}$ is said to have *quantifier elimination* if and only if, for every $n \in \mathbb{N}^+$, every definable subset of A^n is defined by a quantifier-free formula.

Since disjunctions of principal DLO formulas are quantifier-free formulas, Theorem 17.4 implies that $\mathbb{Q}_<$ has quantifier elimination.

17.4 Quantifier Elimination for Theories

Quantifier elimination also makes sense for theories, not just structures. This is a more general notion because it makes sense even for incomplete theories.

Definition 17.6 An L-theory T has *quantifier elimination*, if for every $n \in \mathbb{N}$ (including $n = 0$), for every L-formula $\varphi(x_1, \ldots, x_n)$, there is a quantifier-free L-formula $\theta(x_1, \ldots, x_n)$ such that $T \vdash \forall\bar{x}[\varphi(\bar{x}) \leftrightarrow \theta(\bar{x})]$.

For $n = 0$ in this definition, that is, for sentences, there is a minor technical problem. Consider the sentence $\forall x \exists y[x < y]$. In the theory DLO, this sentence is true, so for DLO to have quantifier elimination we should find a quantifier-free $L_<$-sentence which is a consequence of the DLO-axioms. However, in languages, such as $L_<$, which have no constant symbols, by our original definition of L-formulas, there are no quantifier-free sentences at all! To solve this problem, we introduce two new atomic sentences, $\top$ and $\bot$, which are interpreted as true and false, respectively, in any structure. For a language with a constant symbol c we could use the sentences $c = c$ and $c \neq c$ instead. With this convention we have DLO $\vdash \forall x \exists y[x < y] \leftrightarrow \top$.

Lemma 17.7 *Let $\mathcal{A}$ be an L-structure. Then* Th($\mathcal{A}$) *has quantifier elimination if and only if $\mathcal{A}$ has quantifier elimination.*

Proof Suppose Th($\mathcal{A}$) has quantifier elimination, $n \in \mathbb{N}^+$, and $S \subseteq A^n$ is defined by a formula $\varphi(x_1, \ldots, x_n)$. Then there is a quantifier-free formula $\theta(x_1, \ldots, x_n)$ defining the same set S. So $\mathcal{A}$ has quantifier elimination.

Now suppose that $\mathcal{A}$ has quantifier elimination and $\varphi(x_1,\ldots,x_n)$ is an L-formula. If $n > 0$, then $\varphi(x_1,\ldots,x_n)$ defines a subset of A^n, which is also defined by some quantifier-free formula $\theta(x_1,\ldots,x_n)$. So $\mathcal{A} \models \forall\bar{x}[\varphi(\bar{x}) \leftrightarrow \theta(\bar{x})]$, and hence $\mathrm{Th}(\mathcal{A}) \vdash \forall\bar{x}[\varphi(\bar{x}) \leftrightarrow \theta(\bar{x})]$. If $n = 0$, then φ is a sentence, so it is either true or false in $\mathcal{A}$. So either $\mathrm{Th}(\mathcal{A}) \vdash (\varphi \leftrightarrow \top)$ or $\mathrm{Th}(\mathcal{A}) \vdash (\varphi \leftrightarrow \bot)$. □

Putting together Theorem 17.4 and Lemma 17.7, we have proved the following.

Theorem 17.8 *The theory* DLO *has quantifier elimination.* □

Exercises

17.1 List all of the principal DLO formulas for $n = 3$.

17.2 Prove Lemma 17.2.

17.3 Let $T_{=}^{\infty}$ be the theory of infinite sets in the empty language $L_{=}$. Prove that $T_{=}^{\infty}$ has quantifier elimination.

17.4 Let $\mathcal{A}$ be a finite set considered as an $L_{=}$-structure. Prove that $\mathcal{A}$ has quantifier elimination.

17.5 Show that the empty $L_{=}$-theory does not have quantifier elimination. Observe that, with the previous two exercises, this gives an example of a theory without quantifier elimination, even though all of its models do have quantifier elimination.

17.6 For the language consisting of two constant symbols, c and d, show that the theory T^{∞} stating that there are infinitely many elements has quantifier elimination. By considering the sentence $c = d$, show that the theory has exactly two completions.

17.7 Suppose that L is a language without constant symbols and that T is an L-theory with quantifier elimination. Prove that T is a complete theory.

17.8 Show that the structure $I_{<}$ from Examples 15.2 does not have quantifier elimination. However, show that it does have quantifier elimination if we expand the language by adding constant symbols 0 and 1 for the least and greatest endpoints.

18
Substructure Completeness

In the previous chapter we showed that the theory DLO of dense linear orders without endpoints has quantifier elimination. The method we used relies on the fact that every finite tuple from a dense linear order satisifies a special quantifier-free formula (a principal DLO formula) which we could show determines all the definable sets the tuple lies in. As we shall see in Chapter 25, this method only works because DLO is countably categorical. In this chapter we will give another method for proving quantifier elimination, which also works for theories which are not countably categorical.

18.1 Substructure Completeness and Quantifier Elimination

Consider the substructures of $\mathbb{N}_<$ with domains $\{0, 1\}$ and $\{1, 3\}$. These substructures are both linear orders of length 2 and so are isomorphic to each other as $L_<$-structures. However, they are embedded in $\mathbb{N}_<$ differently in a way which is captured by formulas with quantifiers: for example we have $\mathbb{N}_< \models \exists x[x < 1]$ but $\mathbb{N}_< \models \neg\exists x[x < 0]$. In fact we can see that the pairs $(0, 1)$ and $(1, 3)$ satisfy all the same quantifier-free $L_<$-formulas, but not all the same $L_<$-formulas, and this shows that $\mathbb{N}_<$ does not have quantifier elimination.

More generally, suppose $\bar{a}$ is a tuple from an L-structure $\mathcal{M}$ and $\mathcal{A}$ is the substructure of $\mathcal{M}$ generated by $\bar{a}$, that is, $\mathcal{A}$ consists of the interpretations of all the L-terms $t(\bar{x})$ applied to $\bar{a}$. Then the quantifier-free information about $\bar{a}$ in $\mathcal{M}$ determines the isomorphism type of $\mathcal{A}$. So if $\mathcal{M}$ has quantifier elimination, we would expect this to be all the information there is about how $\mathcal{A}$ embeds as a substructure of $\mathcal{M}$. We now give a theorem which captures this intuition about how quantifier elimination, which is about formulas, relates to substructures of models.

Recall from Chapter 13 that to form the diagram $\mathrm{Diag}(\mathcal{A})$ of an L-structure $\mathcal{A}$, we first expand L to L_A and $\mathcal{A}$ to $\mathcal{A}^+$ by adding constant symbols for each element of A. Then $\mathrm{Diag}(\mathcal{A})$ is the set of all atomic L_A-sentences and negations of atomic L_A-sentences which are true in $\mathcal{A}^+$. Models of $\mathrm{Diag}(\mathcal{A})$ correspond to L-structure extensions of $\mathcal{A}$.

Definition 18.1 An L-theory T is said to be *substructure complete* if, whenever $\mathcal{A}$ is a substructure of a model of T, the theory $T \cup \mathrm{Diag}(\mathcal{A})$ is a complete L_A-theory.

Proposition 18.2 *Let T be an L-theory. Then the following are equivalent.*

(i) *T is substructure complete.*
(ii) *Whenever $\mathcal{A}$ is a finitely generated substructure of a model of T, the theory $T \cup \mathrm{Diag}(\mathcal{A})$ is a complete L_A-theory.*
(iii) *T has quantifier elimination.*

Proof (ii) is just a special case of (i), so (i) $\Rightarrow$ (ii) is immediate.

For (ii) $\Rightarrow$ (iii), let $\varphi(\bar{x})$ be an L-formula, let $\bar{c}$ be new constant symbols, and let $L' = L \cup \{\bar{c}\}$.

Let Σ be the set of all quantifier-free L'-sentences σ such that $T \vdash \varphi(\bar{c}) \to \sigma$.

We claim that $T \cup \Sigma \vdash \varphi(\bar{c})$. Suppose not, for a contradiction. Then there is an L'-structure $\mathcal{M}$ such that $\mathcal{M} \models T \cup \Sigma \cup \neg\varphi(\bar{c})$. Let $\mathcal{A}$ be the substructure of $\mathcal{M}$ generated by $\bar{c}$. Then every element of $\mathcal{A}$ is named by a closed term of L'. So adding new constant symbols for elements of $\mathcal{A}$ does nothing, and we can regard L_A as the same as L', and $\mathrm{Diag}(\mathcal{A})$ as a set of L'-sentences.

Then $\mathcal{M} \models T \cup \mathrm{Diag}(\mathcal{A}) \cup \neg\varphi(\bar{c})$. By assumption, $T \cup \mathrm{Diag}(\mathcal{A})$ is a complete L'-theory, so is equal to $\mathrm{Th}(\mathcal{M})$. So $T \cup \mathrm{Diag}(\mathcal{A}) \vdash \neg\varphi(\bar{c})$.

By compactness, there is a finite subset $\{\psi_1, \ldots, \psi_r\}$ of $\mathrm{Diag}(\mathcal{A})$ such that $T \cup \{\psi_1, \ldots, \psi_r\} \vdash \neg\varphi(\bar{c})$. Let $\psi = \bigwedge_{i=1}^{r} \psi_i$. Then $T \vdash \psi \to \neg\varphi(\bar{c})$. So, taking the contrapositive, $T \vdash \varphi(\bar{c}) \to \neg\psi$. Since $\neg\psi$ is a quantifier-free L'-sentence, $\neg\psi \in \Sigma$.

Now we have $\mathcal{M} \models T \cup \Sigma \cup \mathrm{Diag}(\mathcal{A})$, with $T \cup \mathrm{Diag}(\mathcal{A}) \vdash \psi$ and $\Sigma \vdash \neg\psi$, so $\mathcal{M} \models \psi \wedge \neg\psi$, a contradiction. Thus $T \cup \Sigma \vdash \varphi(\bar{c})$ as claimed.

By compactness again, there is a finite subset Σ_0 of Σ such that $T \cup \Sigma_0 \vdash \varphi(\bar{c})$. Let $\theta(\bar{c}) = \bigwedge \Sigma_0$, with $\theta(\bar{x})$ an L-formula. Then $\theta(\bar{c}) \in \Sigma$ so $T \vdash \varphi(\bar{c}) \leftrightarrow \theta(\bar{c})$.

Since the constant symbols $\bar{c}$ are not in L and T is an L-theory, we deduce $T \vdash \forall \bar{x}[\varphi(\bar{x}) \leftrightarrow \theta(\bar{x})]$. So T has quantifier elimination, as required.

Now to prove (iii) $\Rightarrow$ (i), assume that T has quantifier elimination. Let $\mathcal{M}$ be a model of T and $\mathcal{A}$ a substructure of $\mathcal{M}$. Write $\mathcal{M}_A$ for the expansion of $\mathcal{M}$ by naming each element $a \in A$ by a constant symbol c_a. Let σ be an L_A-sentence. Then σ is of the form $\varphi(c_{a_1}, \ldots, c_{a_n})$ for some L-formula $\varphi(x_1, \ldots, x_n)$. Since T has quantifier elimination, there is a quantifier-free L-formula $\psi(x_1, \ldots, x_n)$ such that

$$T \vdash \forall x_1, \ldots, x_n[\varphi(x_1, \ldots, x_n) \leftrightarrow \psi(x_1, \ldots, x_n)].$$

If $\mathcal{M}_A \models \sigma$, then $\mathcal{M} \models \varphi(\bar{a})$, so $\mathcal{M} \models \psi(\bar{a})$, and so $\mathrm{Diag}(\mathcal{A}) \vdash \psi(c_{a_1}, \ldots, c_{a_n})$. Then $T \cup \mathrm{Diag}(\mathcal{A}) \vdash \sigma$.

Otherwise, $\mathcal{M}_A \not\models \sigma$, so $\mathcal{M}_A \models \neg\sigma$, and similarly, $T \cup \mathrm{Diag}(\mathcal{A}) \vdash \neg\sigma$. So $T \cup \mathrm{Diag}(\mathcal{A})$ is a complete L_A-theory. Thus T is substructure complete. □

18.2 Application to Vector Spaces

Theorem 18.3 *Let K be any field. Then $T^{\infty}_{K\text{-VS}}$ has quantifier elimination.*

Proof We will use Proposition 18.2. Suppose that V is an infinite K-vector space. Let A be a finitely generated $L_{K\text{-VS}}$-substructure of V, that is, a finite dimensional subspace. Let κ be a cardinal larger than $|K|$. We will show that $T^{\infty}_{K\text{-VS}} \cup \mathrm{Diag}(A)$ is κ-categorical. Then, since it has no finite models, by the Łos–Vaught test, it is complete.

So suppose W and W' are models of $T^{\infty}_{K\text{-VS}} \cup \mathrm{Diag}(A)$ of cardinality κ. That is, they are K-vector spaces which contain A as subspaces. Let $\{a_1, \ldots, a_n\}$ be a basis for A, and extend it to bases B of W and B' of W'. By Proposition 14.8, $\dim W = \dim W' = \kappa$, so $|B| = |B'| = \kappa$. Choose a bijection $\alpha : B \to B'$ such that $\alpha(a_i) = a_i$ for $i = 1, \ldots, n$. Then α extends to an isomorphism of K-vector spaces $\pi : W \to W'$. Furthermore, π restricts to the identity map on A, so π is an isomorphism of $L_{K\text{-VS}}(A)$-structures. Thus $T^{\infty}_{K\text{-VS}} \cup \mathrm{Diag}(A)$ is κ-categorical, as required. □

18.3 Quantifier Elimination and Completeness

Substructure completeness does not necessarily imply completeness. From Exercise 17.7, we know that if T is a theory with quantifier elimination in a language with no constant symbols then T is complete. On the other hand, Exercise 17.6 gives an example where this fails in a language with constant symbols. The point is that a complete theory has to specify the truth values of all the quantifier-free sentences. This amounts to stating what the substructure of a model generated by the interpretations of all the constant symbols is, up to isomorphism.

Proposition 18.4 *Suppose an L-theory T has quantifier elimination and there is an L-structure $\mathcal{A}$ which embeds into every model of T. Then T is complete.*

Proof Let $\mathcal{M}_1, \mathcal{M}_2$ be two models of T, and let σ be an L-sentence such that $\mathcal{M}_1 \models \sigma$. Since T has quantifier elimination, there is a quantifier-free L-sentence θ such that $T \vdash (\sigma \leftrightarrow \theta)$. So $\mathcal{M}_1 \models \theta$. By Lemma 5.5, $\mathcal{M}_1 \models \theta$ implies $\mathcal{A} \models \theta$, which implies $\mathcal{M}_2 \models \theta$. So every model of T is a model of θ, and hence $T \vdash \theta$, so $T \vdash \sigma$. Thus $T = \mathrm{Th}(\mathcal{M}_1)$, so T is complete. □

Together, Propositions 18.2 and 17.7 relate completeness and quantifier elimination, giving methods of proving either from the other. For vector spaces we deduced quantifier elimination from completeness.

So far, the only method we have for proving completeness of a theory is the Łos-Vaught test: a theory which is categorical in some infinite cardinality and has no finite models is complete. However, for theories which are not categorical in any infinite cardinality, another method is needed.

A more sophisticated version of the back-and-forth method allows one to prove quantifier elimination by building isomorphisms only between parts of models, not between whole models as is needed for categoricity. Proposition 18.4 can then be used to prove completeness of the theory. It is not essential for the theory to have quantifier elimination in the original language. For example, Exercise 17.8 shows that $\mathrm{DLO}_{\mathrm{ep}}$ has quantifier elimination after adding constant symbols naming the endpoints.

One important application of this extended back-and-forth technique is to the real field.

Fact 18.5 *The theory* RCF *of real closed fields is complete and has quantifier elimination in the language* $L_{\text{o-ring}}$.

The proof is a little beyond the scope of this book. One can also prove that the theory of non-empty discrete linear orders without endpoints is the complete theory of $\mathbb{Z}_<$ using the back-and-forth method. See [Poi00, Section 1.2] for details. We will give a different proof of that theorem in chapter 27.

This more elaborate back-and-forth method is described very elegantly in the first six chapters of Poizat [Poi00] and in Chapter 3 of Hodges [Hod93, Hod97] and is also covered in the books by Marker [Mar02] and Tent and Ziegler [TZ12]. The application to RCF can also be found in these books. The method can be framed as a two-player game, called an Ehrenfeucht–Fräissé game. This approach is stressed in Väänänen [Vää11].

Exercises

18.1 Show that the structure $\mathbb{Z}_<$ does not have quantifier elimination.

18.2 Let L be the language with a single unary relation symbol R. Let T^∞ be the L-theory which says there are infinitely many elements, and for $n \in \mathbb{N}$, let T_n^∞ be the L-theory which also says that there are exactly n elements in the subset named by R. Show that T^∞ is not substructure complete but that each T_n^∞ is.

18.3 Let K be a field. Show that the incomplete theory $T_{K\text{-VS}}$ does not have quantifier elimination but every completion of it does.

18.4 Describe all the finitely generated substructures of models of the theory $T_S = \mathrm{Th}(\mathbb{N}_{\mathrm{succ}})$ from Chapter 16. Prove that T_S is substructure complete. Deduce that $\mathbb{N}_{\mathrm{succ}}$ has quantifier elimination.

18.5 Let $\mathcal{N} = \langle \mathbb{N}; s \rangle$ be the reduct of $\mathbb{N}_{\mathrm{succ}}$ without the constant symbol for 0. Show that $\mathcal{N}$ does not have quantifier elimination.

18.6 Give two isomorphic subrings of $\mathbb{R}_{\mathrm{ring}}$ which illustrate that the real field does not have quantifier elimination in the language L_{ring}.

18.7 Suppose that T is a complete theory with quantifier elimination, that $\mathcal{A}$ and $\mathcal{B}$ are models of T, and that $\pi : \mathcal{A} \to \mathcal{B}$ is an embedding. Show that π is an elementary embedding.

18.8 An L-theory T is said to be *model complete* iff, whenever $\mathcal{M} \models T$, the theory $T \cup \mathrm{Diag}(\mathcal{M})$ is a complete L_M-theory.

(a) Show that T is model complete iff whenever $\mathcal{M}_1, \mathcal{M}_2 \models T$ and $\mathcal{M}_1 \subseteq \mathcal{M}_2$, then $\mathcal{M}_1 \preccurlyeq \mathcal{M}_2$.

(b) Suppose that T 'eliminates quantifiers to the level of existential formulas', that is, for every L-formula $\varphi(\bar{x})$, there is an existential L-formula $\theta(\bar{x})$ such that $T \vdash \forall \bar{x}[\varphi(\bar{x}) \leftrightarrow \theta(\bar{x})]$. Show that T is model complete.

(c) Show that if T is model complete, then it eliminates quantifiers to the level of existential formulas. [Hint: try to adapt the proof of Proposition 18.2.]

19
Power Sets and Boolean Algebras

The notion of a Boolean algebra originally came from logic as a way to capture the logical operations AND, OR, and NOT in an algebraic way. It can also be used to capture the set theoretic operations of intersection, union, and complement. In the next chapter we will see how these two ideas come together when looking at the collection of definable subsets of a structure. In this chapter we introduce Boolean algebras in a slightly roundabout way, by trying to find the complete theory of the power set of an infinite set, considered as a partial order. It is an example of a class of structures which is not first-order axiomatisable, but for which the search for axioms is very fruitful. Even though this is not historically how these axioms were found, it provides another illustration of the process of finding an axiomatisation for the theory of a structure. In particular, it shows how it is often useful to change the language (without changing which sets are definable).

19.1 Power Sets as Partially Ordered Sets

Given a set X, the *power set* $\mathcal{P}X$ of X is the set of all subsets of X. We will investigate the theory of power sets $\mathcal{P}X$ considered as a structure $\langle \mathcal{P}X; \subseteq \rangle$, where $\subseteq$ is the usual subset relation. So the elements of our structure are subsets of X, not elements of X. Let $T_{\text{pow}} = \text{Th}(\{\langle \mathcal{P}X; \subseteq \rangle \mid X \text{ is a set}\})$, and let $T_{\text{pow}}^{\infty} = \text{Th}(\{\langle \mathcal{P}X; \subseteq \rangle \mid X \text{ is an infinite set}\})$. The usual axioms of *partial orders* hold for all power sets:

Transitivity $\forall xyz[(x \subseteq y \wedge y \subseteq z) \rightarrow x \subseteq z]$;

Reflexivity $\forall x[x \subseteq x]$; and

Antisymmetry $\forall xy[(x \subseteq y \wedge y \subseteq x) \rightarrow x = y]$.

However, not every partial order arises from a power set, so we need to consider more properties.

19.2 Lattices

There are natural operations of intersection and union which we can capture. Given $a, b \in \mathcal{P}X$, we usually define $a \cap b$ as $\{x \in X \mid x \in a \wedge x \in b\}$. However, this is not a definition by a formula in our chosen language. Nonetheless, there is a formula which captures the notion of the intersection. We can see that $c = a \cap b$ if and only if $\mathcal{P}X \models \varphi(a, b, c)$, where $\varphi(x_1, x_2, y)$ is the formula

$$y \subseteq x_1 \wedge y \subseteq x_2 \wedge \forall w[(w \subseteq x_1 \wedge w \subseteq x_2) \to w \subseteq y],$$

which says that y is the *greatest lower bound* of x_1 and x_2. So in a power set, every pair of elements has a greatest lower bound. Similarly, it has a *least upper bound*, the union of the two sets. A power set $\mathcal{P}X$ also has a least element, $\emptyset$, and a greatest element, X. So we can add the following axioms to the power set axioms:

GLB $\forall x_1 x_2 \exists y[y \subseteq x_1 \wedge y \subseteq x_2 \wedge \forall w[(w \subseteq x_1 \wedge w \subseteq x_2) \to w \subseteq y]]$;
LUB $\forall x_1 x_2 \exists y[x_1 \subseteq y \wedge x_2 \subseteq y \wedge \forall w[(x_1 \subseteq w \wedge x_2 \subseteq w) \to y \subseteq w]]$;
Least element $\exists y \forall w[y \subseteq w]$; and
Greatest element $\exists y \forall w[w \subseteq y]$.

Any partially ordered set satisfying these axioms is called a *lattice*.

It is convenient to introduce new constant symbols 0 for the least element and 1 for the greatest element. The formula $\varphi(x_1, x_2, y)$ above defines the function $y = x_1 \cap x_2$, so we can use the binary function symbol $\cap$ as an abbreviation, without changing which sets are definable. Similarly, we can use $\cup$ as an abbreviation. Such an expansion of the language, naming a relation, function, or constant which is already definable, is called an *expansion by definitions*.

We could then omit $\subseteq$ from the language, since it is definable from $\cap$ or from $\cup$ (see Exercise 19.3).

We note that $\cap$ and $\cup$ are *commutative* and *associative* in any lattice, and in a power set, they also satisfy the following *distributive laws*:

Distributivity 1 $\forall xyz[x \cap (y \cup z) = (x \cap y) \cup (x \cap z)]$; and
Distributivity 2 $\forall xyz[x \cup (y \cap z) = (x \cup y) \cap (x \cup z)]$.

Not all lattices satisfy these laws, but those which do are called *distributive lattices*.

19.3 Boolean Algebras

Within $\mathcal{P}X$, we can also take the *complement* $a^c = X \setminus a$ of any set a. It can be characterized uniquely by the property that $(a \cup a^c = 1) \wedge (a \cap a^c = 0)$. Not every distributive lattice has complements, so we add the following axiom to our list:

Complementation $\forall x \exists y [x \cap y = 0 \wedge x \cup y = 1]$.

Again, it is convenient to use the function symbol ^{c} as an abbreviation.

Definition 19.1 A *Boolean algebra* is a distributive lattice with complements.

The notion of Boolean algebra is important, but by no means is every Boolean algebra isomorphic to, or even elementarily equivalent to, a power set algebra. For linear orders, there was a natural division into discrete and dense linear orders (although some linear orders are neither). A similar division occurs for Boolean algebras.

19.4 Atoms

A fundamental property of sets is that they are determined by their elements. In set theory this is known as the *axiom of extension*. An element $x \in X$ is not available directly to us, but the singleton set $\{x\}$ is an element of $\mathcal{P}X$. Singleton sets are non-empty, but have no non-empty subsets.

Definition 19.2 An element a of a Boolean algebra $\mathcal{B}$ is called an *atom* if $\mathcal{B} \models \mathrm{At}(a)$, where $\mathrm{At}(x)$ is the formula

$$x \neq 0 \wedge \forall w [w \subsetneq x \rightarrow w = 0].$$

So we can include a version of the axiom of extension in our axioms. In fact, given the other axioms of a Boolean algebra, it is equivalent to something simpler.

Lemma 19.3 *In any Boolean algebra the following two conditions are equivalent:*

Extension property $\forall xyz[(\mathrm{At}(z) \rightarrow (z \subseteq x \leftrightarrow z \subseteq y)) \rightarrow x = y]$; *and*
Atomic $\forall x[x \neq 0 \rightarrow \exists y[\mathrm{At}(y) \wedge y \subseteq x]]$.

Proof Suppose $\mathcal{B}$ satisfies the atomic property, suppose that $b, d \in \mathcal{B}$ have the same set of atoms below them, and let a be an atom of $\mathcal{B}$. If $a \nsubseteq b$, then

$a \not\subseteq b \cap d^c$. But if $a \subseteq b$, then $a \subseteq d$, so $a \not\subseteq d^c$, so $a \not\subseteq b \cap d^c$. So $b \cap d^c$ has no atoms beneath it. By the atomic property, $b \cap d^c = 0$. Similarly, we can show that every atom is beneath $b \cup d^c$. So no atoms are beneath $(b \cup d^c)^c$, so $(b \cup d^c)^c = 0$, so $b \cup d^c = 1$. So d^c satisfies the properties of being a complement of b, but complements are unique, so $d^c = b^c$, and taking complements of both sides, it follows that $d = b$. So the extension property holds. The converse direction is immediate. □

Definition 19.4 A Boolean algebra satisfying the equivalent conditions of Lemma 19.3 is called an *atomic Boolean algebra.*

Fact 19.5 *The axioms for atomic Boolean algebras axiomatise the theory* T_{pow}*, and adding axioms saying the structure is infinite gives a complete axiomatisation of* T^{∞}_{pow}.

The theory T^{∞}_{pow} is not categorical in any infinite cardinal, so we cannot use the Łos–Vaught test to prove completeness. The back-and-forth method can be used as mentioned in Section 18.3. See [Poi00, Section 6.3] for details.

Not every model of T^{∞}_{pow} is isomorphic to a power set algebra, so the class of power set algebras is not an axiomatisable class.

Example 19.6 Let $B = \{a \subseteq \mathbb{N} \mid a \text{ is finite or } \mathbb{N} \setminus a \text{ is finite}\}$, and let $\mathcal{B} = \langle B, \subseteq \rangle$. Then $\mathcal{B}$ is closed in $\mathcal{P}\mathbb{N}$ under the operations $\cap$, $\cup$, and ^{c}, so (using Exercise 19.6, or just by a direct proof) it is a Boolean algebra. Also, if $a \in \mathcal{B}$ with $a \neq 0$, then a is a nonempty set, so has an element, say, s. Then $\{s\} \subseteq a$. So $\mathcal{B}$ is atomic. However, it is countably infinite, and no power set can be countably infinite.

Remarks 19.7 (i) We have considered power sets and Boolean algebras in the language $L_{\subseteq} = \langle \subseteq \rangle$, and we introduced the constants 0 and 1 and the function symbols $\cap$, $\cup$, and ^{c} as abbreviations. However, we could also use the language $L_{\text{Bool}} = \langle \cap, \cup, ^c, 0, 1 \rangle$. In this case the axioms for Boolean algebras will be different. Chapter 2 of Cori and Lascar [CL00] gives a clear account of this and many other topics on Boolean algebras.

(ii) We have used the symbols $\subseteq$, $\cap$, and $\cup$ for the partial order, greatest lower bound, and least upper bound in a lattice or Boolean algebra, respectively, even when it is not a power set algebra. It is more conventional to use $\leqslant$, $\wedge$, and $\vee$, but I have avoided these symbols because the latter two conflict with our symbols for AND and OR, which sometimes appear in the same formulas.

Exercises

19.1 Show that any linearly ordered set with endpoints is a distributive lattice, but that if it has at least three elements, it is not a Boolean algebra.

19.2 Give examples of a partially ordered set which is not a lattice and a lattice which is not distributive.

19.3 Show that in a lattice, the partial order $\subseteq$ can be defined just using $\cap$ or using $\cup$.

19.4 A Boolean algebra $\mathcal{B}$ is *complete* if every subset of $\mathcal{B}$ (not just the finite subsets) has a greatest lower bound. Show that a complete atomic Boolean algebra $\mathcal{B}$ is isomorphic to the power set algebra $\mathcal{P}(\mathrm{At}(\mathcal{B}))$.

19.5 Suppose that $\mathcal{B}$ is a finite Boolean algebra. Show that $\mathcal{B}$ is complete and atomic and that if n is the number of atoms of $\mathcal{B}$, then $|\mathcal{B}| = 2^n$.

19.6 Show that in the language $\langle \cap, \cup, {}^c, 0, 1\rangle$, the theory of Boolean algebras can be axiomatised by $\forall$-sentences. Use Proposition 5.7 to deduce that if $\mathcal{B}$ is a subset of $\mathcal{P}X$ containing $\emptyset$ and X which is closed under $\cap$, $\cup$, and ^{c}, then $\mathcal{B}$ is a Boolean subalgebra of $\mathcal{P}X$.

19.7 A *Boolean ring* is a ring $\langle R; +, \cdot, -, 0, 1\rangle$ which satisfies $\forall x[x \cdot x = x]$. In a Boolean ring, define $x \cap y = x \cdot y$, define $x \cup y = x + y + x \cdot y$, and define $x^c = 1 + x$. Show that $\langle R; \cap, \cup, {}^c, 0, 1\rangle$ is a Boolean algebra. Conversely, if $\mathcal{B}$ is any Boolean algebra, show how to define $\cdot$, $+$, and $-$ to make $\mathcal{B}$ into a Boolean ring.

19.8 An *ideal* of a Boolean algebra $\mathcal{B}$ is a subset $I \subseteq \mathcal{B}$ such that $0 \in I$; if $y \in I$ and $x \subseteq y$, then $x \in I$; and if $x \in I$ and $y \in I$, then $x \cup y \in I$.

(a) Show that an ideal of a Boolean algebra $\mathcal{B}$ is the same thing as an ideal of $\mathcal{B}$ considered as a Boolean ring.
(b) Show that the quotient $\mathcal{B}/I$ is also a Boolean ring.

19.9 A Boolean algebra is *atomless* if it satisfies $\neg\exists x \, \mathrm{At}(x)$ and $0 \neq 1$.

(a) Show that any atomless Boolean algebra is infinite.
(b) Let $\mathcal{B}$ be an infinite Boolean algebra, and let $\mathrm{Fin}(\mathcal{B})$ be the ideal generated by $\mathrm{At}(\mathcal{B})$. Show that $\mathcal{B}/\mathrm{Fin}(\mathcal{B})$ is an atomless Boolean algebra.
(c) Use the back-and-forth method to show that the theory of atomless Boolean algebras is countably categorical and has quantifier elimination in the language $\langle \cap, \cup, {}^c, 0, 1\rangle$.

20

The Algebras of Definable Sets

In this chapter we will look more systematically at the definable sets in a model $\mathcal{A}$ of a theory T. The definable subsets of A^n form a Boolean algebra, and we study this in the examples of dense linear orders and vector spaces. We show that the atoms of the Boolean algebras are related to certain formulas, called principal formulas.

20.1 Lindenbaum Algebras

Definition 20.1 Let $\mathcal{A}$ be an L-structure. As usual, A denotes the domain of $\mathcal{A}$. We write $\mathrm{Def}_n(\mathcal{A})$ for the set of all definable subsets of A^n.

Recall that A^n itself and $\emptyset$ are definable and that $\mathrm{Def}_n(\mathcal{A})$ is closed under the Boolean operations $\cap$, $\cup$, and complement. So using Exercise 19.6, it can be considered as a Boolean algebra.

Now we start from the syntactic direction: formulas and theories. Given a language L, we have written $\mathrm{Form}(L)$ for the set of all L-formulas. Now define $\mathrm{Form}_n(L)$ to be the set of all L-formulas with free variables from $x_1, \ldots, x_n$ only. Let T be an L-theory. We define a relation $\sim_T$ on $\mathrm{Form}_n(L)$ by

$$\varphi(\bar{x}) \sim_T \theta(\bar{x}) \quad \text{if and only if} \quad T \vdash \forall \bar{x}[\varphi(\bar{x}) \leftrightarrow \theta(\bar{x})].$$

It is easy to see that $\sim_T$ is an equivalence relation on $\mathrm{Form}_n(L)$. We write $[\varphi(\bar{x})]$ for the equivalence class of $\varphi(\bar{x})$, and we write $\mathrm{Lind}_n(T)$ for the set of equivalence classes.

Lemma 20.2 *The logical operations $\wedge$, $\vee$, and $\neg$ on formulas induce the structure of a Boolean algebra on* $\mathrm{Lind}_n(T)$*. That is, if we define $[\varphi(\bar{x})] \wedge [\theta(\bar{x})]$ to be $[(\varphi \wedge \theta)(\bar{x})]$, and if we define $\vee$ and $\neg$ similarly, we get well-defined operations on* $\mathrm{Lind}_n(T)$ *which make it into a Boolean algebra.*

We leave the proof as an exercise. The algebras $\mathrm{Lind}_n(T)$ are called the *Lindenbaum algebras* of T.

Lemma 20.3 *Let $\mathcal{A}$ be an L-structure and $T = \mathrm{Th}(\mathcal{A})$. There is an isomorphism of Boolean algebras between* $\mathrm{Lind}_n(T)$ *and* $\mathrm{Def}_n(\mathcal{A})$ *given by*

$$[\varphi(\bar{x})] \mapsto \varphi(\mathcal{A}).$$

Again, we leave the proof as an exercise. An important consequence of this lemma is that if $\mathcal{A}$ and $\mathcal{B}$ are models of the same complete theory T, the Boolean algebras $\mathrm{Def}_n(\mathcal{A})$ and $\mathrm{Def}_n(\mathcal{B})$ are isomorphic to each other, because they are both isomorphic to $\mathrm{Lind}_n(T)$, so as abstract Boolean algebras, they do not depend on the choice of model.

20.2 Example: Dense Linear Orders

We take $T = \mathrm{DLO}$ and consider the model $\mathbb{Q}_<$. By Theorem 17.8, DLO has quantifier elimination, so we only need to consider the quantifier-free definable sets. More specifically, Theorem 17.4 shows that every definable set is defined by a finite disjunction of principal DLO formulas.

For example, for $n = 1$, the only principal DLO formula is $x_1 = x_1$, so the definable subsets of $\mathbb{Q}$ are $\mathbb{Q}$ itself, defined by $x_1 = x_1$, and $\emptyset$, defined by the empty disjunction, which is equivalent to the negation $x_1 \neq x_1$. For $n = 2$, the principal DLO formulas are $x_1 = x_2$, $x_2 = x_1$, $x_1 < x_2$, and $x_2 < x_1$. The first two define the same subset of $\mathbb{Q}^2$, so there are three different sets defined by principal DLO formulas. Then there are $2^3 = 8$ different definable subsets of $\mathbb{Q}^2$ corresponding to the different possible disjunctions of these sets. In general, if the principal DLO formulas for a given n define m different subsets of $\mathbb{Q}^n$, then the Boolean algebra $\mathrm{Def}_n(\mathbb{Q}_<)$ has size 2^m and is isomorphic to the power set algebra on a finite set of size m.

20.3 Principal Formulas

There are analogues of these principal DLO formulas in other theories.

Definition 20.4 Let T be a complete L-theory. A *principal formula* (with respect to T) is a formula $\psi(\bar{x})$ such that

(i) $T \vdash \exists\bar{x}[\psi(\bar{x})]$, (we say $\psi(\bar{x})$ is *satisfiable)*, and
(ii) for every L-formula $\varphi(\bar{x})$ with the same list of variables, either

$$T \vdash \forall \bar{x}[\psi(\bar{x}) \to \varphi(\bar{x})] \text{ or } T \vdash \forall \bar{x}[\psi(\bar{x}) \to \neg\varphi(\bar{x})].$$

So principal DLO formulas are principal formulas for the theory DLO, and every principal formula for DLO is $\sim_{\text{DLO}}$-equivalent to a principal DLO formula.

For any L-structure $\mathcal{A}$, a formula $\varphi(\bar{x}) \in \text{Form}_n(L)$ is a principal formula for $\text{Th}(\mathcal{A})$ if and only if $\varphi(\mathcal{A})$ is an atom of $\text{Def}_n(\mathcal{A})$. So to understand $\text{Def}_n(\mathcal{A})$, a good start is to try to understand all the principal formulas. When there are only finitely many, and every n-tuple from $\mathcal{A}$ satisfies one of them, this is enough to determine all the definable sets. The following proposition captures the method we used for DLO.

Proposition 20.5 *Suppose $\psi_1(\bar{x}), \ldots, \psi_m(\bar{x})$ are all principal formulas in n variables for the complete theory $T = \text{Th}(\mathcal{A})$, defining different subsets of A^n, and $\bigcup_{i=1}^m \psi_i(\mathcal{A}) = A^n$. Then every definable set in $\text{Def}_n(\mathcal{A})$ is a union of some of the sets $\psi_i(\mathcal{A})$, and $|\text{Def}_n(\mathcal{A})| = 2^m$.* □

If there are only finitely many definable subsets of A^n, then they must follow this pattern.

Lemma 20.6 *If $\text{Def}_n(\mathcal{A})$ is finite, then every $\bar{a} \in A^n$ satisfies a principal formula.*

Proof Given $\bar{a} \in A^n$, if $\text{Def}_n(\mathcal{A})$ is finite, then there is a minimal definable subset containing $\bar{a}$, and any formula defining it is a principal formula. □

The converse of this lemma is false. For example, in $\text{Def}_1(\mathbb{N}_{\text{s-ring}})$, every $m \in \mathbb{N}$ satisfies a principal formula $x_1 = 0$ or $x_1 = \underbrace{1 + \cdots + 1}_{m}$, but $\text{Def}_1(\mathbb{N})$ is infinite.

20.4 Vector Spaces over Finite Fields

Recall the theory $T^{\infty}_{\mathbb{F}_3\text{-VS}}$ of infinite vector spaces over the field $\mathbb{F}_3$ with three elements: 0, 1, and 2. Let $V \models T^{\infty}_{\mathbb{F}_3\text{-VS}}$. From Theorem 18.3 we know that $T^{\infty}_{\mathbb{F}_3\text{-VS}}$ has quantifier elimination. So every definable set is defined by a Boolean combination of atomic formulas. The atomic formulas in variables $x_1, \ldots, x_n$ are all equivalent under the theory $T^{\infty}_{\mathbb{F}_3\text{-VS}}$ to formulas of the form $\sum_{i=1}^n \lambda_i x_i = 0$ for scalars $\lambda_i \in \mathbb{F}_3$.

For $n = 1$, this gives us atomic formulas $0 = 0$, $x_1 = 0$, and $2x_1 = 0$. The latter two both define $\{0\}$, which is a singleton, so $x_1 = 0$ is a principal formula (see Exercise 20.6). The atomic formula $0 = 0$ defines all of V and is not a principal formula. However, the negation $x_1 \neq 0$ is principal, which we can see because no atomic formula splits up the set it defines into proper subsets. Thus $\mathrm{Def}_1(V)$ consists of exactly the four sets $\emptyset$, $\{0\}$, $V \setminus \{0\}$, and V.

$\mathrm{Def}_2(V)$ is a bit more interesting, because for each $\lambda \in \mathbb{F}_3$, we have the formula $x_2 = \lambda \cdot x_1$, giving a linear equation. Suppose $\lambda \in \mathbb{F}_3$ and we have $(a_1, a_2) \in V^2$ and $(b_1, b_2) \in V^2$ such that

$$V \models a_1 \neq 0 \wedge a_2 = \lambda \cdot a_1 \quad \text{and also} \quad V \models b_1 \neq 0 \wedge b_2 = \lambda \cdot b_1.$$

Then there is an automorphism π of V such that $\pi(a_1, a_2) = (b_1, b_2)$. So the subset of V^2 defined by the formula $x_1 \neq 0 \wedge x_2 = \lambda \cdot x_1$ has no proper non-empty definable subsets, and hence the formula is principal. (Here we have used automorphisms directly, as in the proof of the quantifier elimination theorem. We could also have used the theorem.)

We have the following five formulas in two variables:

$$x_1 = 0 \wedge x_2 = 0, \qquad x_1 = 0 \wedge x_2 \neq 0,$$

$$x_1 \neq 0 \wedge x_2 = 0, \qquad x_1 \neq 0 \wedge x_2 = x_1, \qquad x_1 \neq 0 \wedge x_2 = 2 \cdot x_1,$$

which are all principal. If (a_1, a_2) satisfies one of these formulas, then a_1 and a_2 are linearly dependent, so there is also the possibility that (a_1, a_2) is a linearly independent pair of vectors. We can also capture this situation with a single formula,

$$\bigwedge_{(\lambda_1,\lambda_2) \in \mathbb{F}_3^2 \setminus \{(0,0)\}} \lambda_1 \cdot x_1 + \lambda_2 \cdot x_2 \neq 0,$$

which is first order because it is a finite conjunction. Every linearly independent pair of elements of V satisfies the same atomic formulas, hence the same quantifier-free formulas, hence (by quantifier elimination) the same formulas. So this is again a principal formula.

So every pair of elements from V satisfies exactly one of these six principal formulas. Hence $\mathrm{Def}_2(V)$ has exactly $2^6 = 64$ definable sets.

Exercises

20.1 Prove Lemma 20.2.

20.2 Prove Lemma 20.3.

20.3 Suppose that T is an incomplete theory and $\mathcal{A} \models T$. Show there is a surjective homomorphism of Boolean algebras $\mathrm{Lind}_n(T) \to \mathrm{Def}_n(\mathcal{A})$ as in Lemma 20.3, but that it is not injective.

20.4 Give two examples of principal formulas and two examples of formulas which are satisfiable but non-principal in the theory of $\mathbb{Z}_{\text{ring}}$.

20.5 How many definable sets are there in $\text{Def}_n(\mathbb{Q}_<)$ for $n = 3$ and $n = 4$?

20.6 Suppose that $\varphi(x)$ is a formula such that $T \vdash \exists^{=1} x[\varphi(x)]$. Show that $\varphi(x)$ is a principal formula for T.

20.7 Consider the expansion $\mathcal{A}$ of $\mathbb{Q}_<$ by constant symbols naming 0 and 1. How many definable sets are there in $\text{Def}_n(\mathcal{A})$ for $n = 1, 2$?

20.8 How many definable sets are there in $\text{Def}_n(I_<)$ for $n = 1, 2$, where I is the closed unit interval $[0, 1]$ in the reals?

20.9 Let V be an infinite-dimensional $\mathbb{F}_3$-vector space. Show that any pair $(a_1, a_2) \in V^2$ satisfies one of the six principal formulas given before Definition 20.4.

20.10 How many definable sets are there in $\text{Def}_n(V)$ for $n = 1, 2, 3$, where V is an infinite $\mathbb{F}_4$-vector space?

20.11 Let $\mathcal{A} = \langle \mathbb{Z}; <, 0 \rangle$. Show that for every $n \in \mathbb{N}^+$, $\text{Def}_n(\mathcal{A})$ is an atomic Boolean algebra which is not complete.

20.12 Since Lindenbaum algebras are Boolean algebras, they satisfy the distributive laws and de Morgan laws, which allow us to swap the order of conjunctions, disjunctions, and negations. This allows us to give normal forms for formulas which can be useful in understanding definable sets. Let T be any L-theory, for example the empty L-theory. Let ψ, φ_i, and φ_{ij} be any L-formulas for $i = 1, \ldots, r$ and $j = 1, \ldots, s_i$.

(a) Prove the following logical equivalences:

(1) $\psi \wedge \bigvee_{i=1}^{r} \varphi_i \sim_T \bigvee_{i=1}^{r} (\psi \wedge \varphi_i)$.
(2) $\psi \vee \bigwedge_{i=1}^{r} \varphi_i \sim_T \bigwedge_{i=1}^{r} (\psi \vee \varphi_i)$.
(3) $\bigwedge_{i=1}^{r} \bigvee_{j=1}^{s_i} \varphi_{ij} \sim_T \bigvee_{j_1=1}^{s_1} \cdots \bigvee_{j_r=1}^{s_r} \bigwedge_{i=1}^{r} \varphi_{ij_i}$.
(4) $\bigvee_{i=1}^{r} \bigwedge_{j=1}^{s_i} \varphi_{ij} \sim_T \bigwedge_{j_1=1}^{s_1} \cdots \bigwedge_{j_r=1}^{s_r} \bigvee_{i=1}^{r} \varphi_{ij_i}$.
(5) $\neg \bigwedge_{i=1}^{r} \varphi_i \sim_T \bigvee_{i=1}^{r} \neg\varphi_i$.
(6) $\neg \bigvee_{i=1}^{r} \varphi_i \sim_T \bigwedge_{i=1}^{r} \neg\varphi_i$.

(b) Conjunctive normal form theorem: suppose that φ is a quantifier-free formula. Prove by induction on the construction of φ that there is an L-formula θ such that $\theta \sim_T \varphi$, and θ is of the form $\bigwedge_{i=1}^{r} \bigvee_{j=1}^{s_i} \alpha_{ij}$, where each α_{ij} is either an atomic formula or the negation of an atomic formula.

(c) Disjunctive normal form theorem: show that every quantifier-free formula φ is also logically equivalent to a formula ψ of the form

$\bigvee_{i=1}^{r} \bigwedge_{j=1}^{s_i} \beta_{ij}$, where each β_{ij} is either an atomic formula or the negation of an atomic formula.

(d) Show that if φ is a *positive* quantifier-free formula, that is, it is constructed using $\wedge$ and $\vee$, but without implications or negations, then its conjunctive and disjunctive normal forms also do not use negation.

21

Real Vector Spaces and Parameters

In the previous chapter we characterised the Lindenbaum algebras of definable sets for DLO and for vector spaces over finite fields, where these algebras are finite. In this chapter we consider vector spaces over the real field where the algebras are not finite. The pattern is the same for any infinite field. We also consider the sets which become definable when new constants, called parameters, are added to the language.

21.1 The Real Line as a Vector Space

Consider the real line $\mathbb{R}$ as a one-dimensional $\mathbb{R}$-vector space. We write it as $\mathbb{R}_{\mathbb{R}\text{-VS}}$. By Theorem 18.3, $\mathbb{R}_{\mathbb{R}\text{-VS}}$ has quantifier elimination. Atomic formulas in the variables $x_1, \ldots, x_n$ are equivalent to formulas of the form $\sum_{i=1}^{n} \lambda_i x_i = 0$ for scalars $\lambda_i \in \mathbb{R}$. Conjunctions of atomic formulas therefore define vector subspaces of $\mathbb{R}^n$, and indeed all subspaces of $\mathbb{R}^n$ are definable. Thus $\mathrm{Def}_n(\mathbb{R}_{\mathbb{R}\text{-VS}})$ consists of all the Boolean combinations of subspaces of $\mathbb{R}^n$.

The only subspace of $\mathbb{R}^1$ is $\{0\}$, so $\mathrm{Def}_1(\mathbb{R}_{\mathbb{R}\text{-VS}})$ consists of the four sets $\emptyset$, $\{0\}$, $\mathbb{R} \smallsetminus \{0\}$, and $\mathbb{R}$.

The subspaces of $\mathbb{R}^2$ are $\{(0,0)\}$, $\mathbb{R}^2$ itself, and the one-dimensional subspaces given by $x_2 = \lambda x_1$ for each $\lambda \in \mathbb{R}$, and by $x_1 = 0$.

Lemma 21.1 *For each $n \in \mathbb{N}^+$, the formula $\bigwedge_{i=1}^{n} x_i = 0$ defining the origin 0 is a principal formula. If L is any 1-dimensional subspace of $\mathbb{R}^n$, then any formula defining $L \smallsetminus \{0\}$ is a principal formula.*

Proof It is immediate that $\bigwedge_{i=1}^{n} x_i = 0$ is principal since it defines a singleton set. Suppose that L is a one-dimensional subspace of $\mathbb{R}^n$ and $a, b \in L \smallsetminus \{0\}$. Then there is $\lambda \in \mathbb{R} \smallsetminus \{0\}$ such that $b = \lambda a$. Multiplication by λ is an automorphism of

$\mathbb{R}_{\mathbb{R}\text{-VS}}$, so a, b must satisfy the same formulas. So any formula defining $L \setminus \{0\}$ is principal. □

Proposition 21.2 *For all* $n \in \mathbb{N}^+$, $\mathrm{Def}_n(\mathbb{R}_{\mathbb{R}\text{-VS}})$ *is an atomic Boolean algebra. It is not complete if* $n \geqslant 2$.

Proof Let $S \subseteq \mathbb{R}^n$ be definable and non-empty. We must show that there is an atom of $\mathrm{Def}_n(\mathbb{R}_{\mathbb{R}\text{-VS}})$ contained in S. Let $a = (a_1, \ldots, a_n) \in S$. If $a = 0$, then $\{0\} \subseteq S$. Otherwise, there is i such that $a_i \neq 0$. For convenience, assume $a_1 \neq 0$. Then for each $j = 2, \ldots, n$, there is $\lambda_j \in \mathbb{R}$ such that $a_j = \lambda_j a_1$. Let $\varphi(\bar{x})$ be the formula $x_1 \neq 0 \wedge \bigwedge_{j=2}^{n} x_j = \lambda_j x_1$. Then $\varphi(\bar{x})$ is a principal formula by Lemma 21.1, and the set it defines is contained in S. So in either case, S contains an atom of the Boolean algebra $\mathrm{Def}_n(\mathbb{R}_{\mathbb{R}\text{-VS}})$, so $\mathrm{Def}_n(\mathbb{R}_{\mathbb{R}\text{-VS}})$ is atomic.

If $n \geqslant 2$, then the number of atoms in $\mathrm{Def}_n(\mathbb{R}_{\mathbb{R}\text{-VS}})$ is $|\mathbb{R}|$, but the total number of formulas is also $|\mathbb{R}|$, so $|\mathrm{Def}_n(\mathbb{R}_{\mathbb{R}\text{-VS}})| = |\mathbb{R}|$. However, a complete atomic Boolean algebra with $|\mathbb{R}|$ atoms has size $|\mathcal{P}\mathbb{R}|$, which is strictly larger than $|\mathbb{R}|$. □

So as in the case of vector spaces over finite fields, principal formulas are an important starting point in determining the definable sets. However, not every definable set is a finite disjunction of these sets; indeed, $\mathbb{R}^2$ is not. So while principal formulas are still important, they do not give the whole picture.

Now suppose that V is another model of $T^{\infty}_{\mathbb{R}\text{-VS}}$, for example $V = \mathbb{R}^3$. For each n we have $\mathrm{Def}_n(V) \cong \mathrm{Lind}_n(T_{\mathbb{R}\text{-VS}}) \cong \mathrm{Def}_n(\mathbb{R}_{\mathbb{R}\text{-VS}})$, so the definable subsets of V^n are not Boolean combinations of all the vector subspaces of $V^n = \mathbb{R}^{3n}$ but only Boolean combinations of certain subspaces. This is because the variables x_i range over the vectors in V and not over the real numbers. In the proof of Proposition 21.2, to find the scalars λ, we made essential use of the fact that we were working in the one-dimensional model $\mathbb{R}_{\mathbb{R}\text{-VS}}$ of $T_{\mathbb{R}\text{-VS}}$. However, the conclusion of the proposition also holds for $\mathrm{Def}_n(V)$. This is an example where we get a benefit from working in a particular model of the theory.

21.2 Parameters

We have seen that definable sets in $\mathbb{R}_{\mathbb{R}\text{-VS}}$ correspond to Boolean combinations of homogeneous linear equations, that is, equations which state that a linear combination of the variables is equal to 0. Often in linear algebra, one wishes to consider systems of inhomogeneous equations, that is, equations of the form $\sum_{i=1}^{n} \lambda_i x_i = a$, for some $a \neq 0$. We can do this by adding a constant symbol to the language for a. In this case, constants representing elements of the structure are called *parameters*.

Definition 21.3 Let $\mathcal{M}$ be an L-structure and A a subset of the domain M of $\mathcal{M}$. A subset $S \subseteq M^n$ is said to be *definable with parameters from A*, or *A-definable*, if there is an L-formula $\varphi(x_1, \ldots, x_n, y_1, \ldots, y_m)$ and elements $a_1, \ldots, a_m \in A$ such that

$$S = \{(x_1, \ldots, x_n) \in M^n \mid \mathcal{M} \models \varphi(x_1, \ldots, x_n, a_1, \ldots, a_m)\}.$$

If $A = M$, then S is said to be *definable with parameters*, or *parametrically definable*. If $A = \emptyset$, then S is sometimes said to be *$\emptyset$-definable*, *null-definable*, *zero-definable*, or *definable without parameters*.

In principle, there is nothing very new here. We can expand the language L to L_A and expand the structure $\mathcal{M}$ to $\mathcal{M}_A$ by adding constants naming each element of A. Then $\mathrm{Th}_{L_A}(\mathcal{M}_A)$ is a complete L_A-theory, and the subsets of M^n which are A-definable with respect to the language L and the same as the subsets which are $\emptyset$-definable with respect to the language L_A. However, in practice, it is very useful to consider $\mathcal{M}$ and $\mathcal{M}_A$ not as totally different structures, because they are much more closely related than the structures you can get by changing the language in other ways. For example, quantifier elimination is preserved.

Lemma 21.4 *Suppose $\mathcal{M}$ is an L-structure with quantifier elimination and $S \subseteq M^n$ is a subset definable with parameters. Then S is defined using the same parameters by a quantifier-free L-formula.*

Proof Suppose $S = \{\bar{x} \in M^n \mid \mathcal{M} \models \varphi(\bar{x}, \bar{a})\}$. By quantifier elimination for $\mathcal{M}$, there is a quantifier-free L-formula $\theta(\bar{x}, \bar{y})$ such that $\mathcal{M} \models \forall \bar{x}\bar{y}[\theta(\bar{x}, \bar{y}) \leftrightarrow \varphi(\bar{x}, \bar{y})]$. So $\theta(\bar{x}, \bar{a})$ is a quantifier-free formula defining S with the same parameters. □

Consider now the parametrically definable subsets of $\mathbb{R}^2$ with respect to the structure $\mathbb{R}_{\mathbb{R}\text{-VS}}$. As well as the origin $(0, 0)$, any point (a, b) in $\mathbb{R}^2$ is now definable by $x_1 = a \wedge x_2 = b$. As well as the straight lines through the origin, formulas of the form $\lambda_1 x_2 + \lambda_2 x_2 = a$ with $\lambda_1, \lambda_2, a \in \mathbb{R}$ now give all the straight lines in $\mathbb{R}^2$. So the parametrically definable subsets of $\mathbb{R}^2$ are the Boolean combinations of points and straight lines.

Exercises

21.1 Let $V = \mathbb{R}^2$, considered as an $\mathbb{R}$-vector space. Then both $\mathrm{Def}_2(V)$ and $\mathrm{Def}_4(\mathbb{R}_{\mathbb{R}\text{-VS}})$ are Boolean algebras of subsets of $\mathbb{R}^4$. Give an example of a subset $S \subseteq \mathbb{R}^4$ which is in $\mathrm{Def}_4(\mathbb{R}_{\mathbb{R}\text{-VS}})$ but not in $\mathrm{Def}_2(V)$.

21.2 Let $S \in \mathrm{Def}_2(\mathbb{R}_{\mathbb{R}\text{-VS}})$. Show that either S or $\mathbb{R}^2 \setminus S$ is defined by a finite disjunction of principal formulas. Is the same true for every $S \in \mathrm{Def}_3(\mathbb{R}_{\mathbb{R}\text{-VS}})$?

21.3 Let $\mathcal{A}$ be the structure $\mathbb{R}_{\mathbb{R}\text{-VS}}$ expanded by a single constant symbol naming 1. Find the atoms in the Lindenbaum algebra $\mathrm{Def}_2(\mathcal{A})$.

21.4 Let $\mathcal{M}$ be an L-structure, suppose $A \subseteq M$ is a set of parameters, and let $n \in \mathbb{N}$. Show that $\mathrm{Def}_n(\mathcal{M}_A) = \bigcup\{\mathrm{Def}_n(\mathcal{M}_{A_0}) \mid A_0$ is a finite subset of $A\}$.

21.5 Let $\mathcal{M}$ be an infinite set considered as an $L_=$=structure, and let $A \subseteq M$. How large is $\mathrm{Def}_n(\mathcal{M}_A)$ for $n = 1, 2$, when $|A| = r$, finite, and when $|A| = \aleph_0$?

21.6 Let $\mathcal{M}$ be an L-structure and $A \subseteq M$ a subset. Recall that the *substructure of $\mathcal{M}$ generated by A* is the smallest L-substructure of $\mathcal{M}$ containing A. Write it as $\langle A \rangle$. Show that $\mathrm{Def}_n(\mathcal{M}_A) = \mathrm{Def}_n(\mathcal{M}_{\langle A \rangle})$.

21.7 The *definable closure* of a subset A of an L-structure $\mathcal{M}$, written $\mathrm{dcl}(A)$, is the set of all $b \in M$ such that the singleton set $\{b\}$ is definable in $\mathcal{M}$, with parameters from A. Show that $\langle A \rangle \subseteq \mathrm{dcl}(A)$ and that $\mathrm{Def}_n(\mathcal{M}_A) = \mathrm{Def}_n(\mathcal{M}_{\mathrm{dcl}(A)})$.

21.8 Suppose that $S \subseteq \mathbb{R}$ is parametrically definable with respect to the structure $\mathbb{R}_{\mathbb{R}\text{-VS}}$. Show that either S or $\mathbb{R} \setminus S$ is a finite set. In the latter case, we say that S is a *cofinite* subset. Conversely, show that every finite subset and every cofinite subset of $\mathbb{R}$ is parametrically definable with respect to $\mathbb{R}_{\mathbb{R}\text{-VS}}$.

21.9 The Lindenbaum algebras of structures are not always atomic. Here is one example. Let L be the language with unary relation symbols D_n for $n \in \mathbb{N}^+$, and consider the structure $\mathcal{R}$ with domain $\mathbb{R}$ and with D_n interpreted as the set of real numbers whose n^{th} decimal digit after the decimal point is even. Show that $\mathcal{R}$ has quantifier elimination by showing that if $r, s \in \mathbb{R}$ satisfy the same quantifier-free formulas, then there is an automorphism of $\mathcal{R}$ sending r to s. Then show that $\mathrm{Def}_1(\mathcal{R})$ is atomless.

21.10 Recall $T_S = \mathrm{Th}(\mathbb{N}_{\mathrm{succ}})$ from Chapter 16. Using the methods of that chapter or using Exercise 18.4, show that if $\mathcal{M}$ is any model of T_S, then every subset of M definable with parameters is finite or cofinite. Show also that no linear order is definable on $\mathcal{M}$, even using parameters.

22
Semi-algebraic Sets

We consider now the subsets of $\mathbb{R}^n$ which are definable with parameters, with respect to the structure $\mathbb{R}_{\text{o-ring}}$. These subsets are also known as *semi-algebraic sets*. As with many structures, the study of the definable sets combines model-theoretic ideas with ideas from the branch of mathematics whence the structure naturally comes, which in this case is elementary real analysis. For simplicity, we will concentrate on the definable subsets of $\mathbb{R}$ and $\mathbb{R}^2$, although most of the ideas needed in higher dimensions are already present here.

22.1 O-Minimality

We will use Fact 18.5, which states that $\mathbb{R}_{\text{o-ring}}$ has quantifier elimination in the ordered ring language. So, as in the previous chapter, we start by considering the sets defined by atomic formulas, and then by quantifier-free formulas.

Lemma 22.1 *Every atomic $L_{\text{o-ring}}$-formula, with parameters from $\mathbb{R}$, defines the same subset of $\mathbb{R}^n$ as a formula of the form $f(\bar{x}) = 0$ or $f(\bar{x}) > 0$, where $f(\bar{x})$ is a polynomial with real coefficients.*

The proof is left as an exercise.

Lemma 22.2 *Every subset of $\mathbb{R}^n$ defined by a quantifier-free $L_{\text{o-ring}}$-formula, with parameters from $\mathbb{R}$, is a finite union of sets defined by formulas of the form $h(\bar{x}) = 0 \wedge \bigwedge_{j=1}^{s} g_j(\bar{x}) > 0$.*

Proof Suppose S is defined by a quantifier-free formula $\varphi(\bar{x})$. Then $\varphi(\bar{x})$ is a Boolean combination of atomic formulas. By the disjunctive normal form theorem, Exercise 20.12, we can write $\varphi(\bar{x})$ such that negation is applied only to atomic formulas. But for any polynomial $f(\bar{x})$,

$$\mathbb{R}_{\text{o-ring}} \models \forall \bar{x}[\neg(f(\bar{x}) = 0) \leftrightarrow (f(\bar{x}) > 0 \vee -f(\bar{x}) > 0)]$$

and

$$\mathbb{R}_{\text{o-ring}} \models \forall \bar{x}[\neg(f(\bar{x}) > 0) \leftrightarrow (f(\bar{x}) = 0 \vee -f(\bar{x}) > 0)].$$

So S is defined by a positive Boolean combination of atomic formulas. Using the disjunctive normal form theorem again, S is defined by a formula of the form

$$\bigvee_{i=1}^{r} \left(\bigwedge_{j=1}^{t_i} h_{ij}(\bar{x}) = 0 \wedge \bigwedge_{j=1}^{s_i} g_{ij}(\bar{x}) > 0 \right).$$

Let $h_i(\bar{x}) = \sum_{j=1}^{t_i} h_{ij}(\bar{x})^2$. Then S is defined by

$$\bigvee_{i=1}^{r} \left(h_i(\bar{x}) = 0 \wedge \bigwedge_{j=1}^{s_i} g_{ij}(\bar{x}) > 0 \right),$$

so it is a finite union of sets of the desired form. □

Now we restrict to the case $n = 1$. A formula of the form $f(x) = 0$ defines a finite set unless f is a constant polynomial when it defines either all of $\mathbb{R}$ or $\emptyset$. For example, $x^3 - 2x = 0$ defines $\{-\sqrt{2}, 0, \sqrt{2}\}$.

We can work out what set is defined by a formula of the form $f(x) > 0$ by considering how the sign of $f(x)$ changes between its zeros. For example, $x^3 - 2x > 0$ defines the set $(-\sqrt{2}, 0) \cup (\sqrt{2}, \infty)$. In general, any finite union of open intervals can be obtained this way. By combining these intervals with points, we can get closed intervals and half-open intervals. Using quantifier elimination and Lemma 22.2, we get the following.

Proposition 22.3 *Every subset of $\mathbb{R}$ which is parametrically definable with respect to the structure $\mathbb{R}_{\text{o-ring}}$ is a finite union of points and intervals.*

Note that these subsets of $\mathbb{R}$ are also parametrically definable in the reduct $\mathbb{R}_{<}$, although not necessarily with the same parameters. However, there are definable sets in two or more free variables, such as that defined by $y = x^2$, which are not definable in $\mathbb{R}_{<}$.

The conclusion of Proposition 22.3 is a very useful condition that has been shown to hold in many other structures and has been given a name.

Definition 22.4 An infinite structure $\mathcal{M}$ in a language containing $<$ is said to be *o-minimal* if M is densely linearly ordered by $<$ and every parametrically definable subset of M is a finite union of points and intervals.

22.2 Continuity of Definable Functions

In o-minimal structures, the study of definable sets goes hand in hand with the study of the definable functions. The parametrically definable functions in $\mathbb{R}_{\text{o-ring}}$ are called *semi-algebraic functions*. Polynomial functions are semi-algebraic, but there are also other semi-algebraic functions.

Example 22.5 The function $f : \mathbb{R} \to \mathbb{R}$ in Figure 22.1 is defined by the formula

$$(x < -2 \wedge (y+1)^3 = x+3) \vee (y \leqslant 0 \wedge (x+1)^2 + y^2 = 1) \vee (0 < x \wedge y^2 = x).$$

This function is not a polynomial function, but it is at least continuous. Not every definable function is continuous; for example the function

$$g(x) = \begin{cases} 0 \text{ if } x < 0 \\ 1 \text{ if } x \geqslant 0 \end{cases}$$

is not continuous at 0, but it is continuous everywhere else in $\mathbb{R}$. O-minimality at once implies that the function

$$h(x) = \begin{cases} 0 \text{ if } x \in \mathbb{Q} \\ 1 \text{ if } x \notin \mathbb{Q} \end{cases}$$

is not definable in $\mathbb{R}_{\text{o-ring}}$. In fact, definable functions are piecewise continuous.

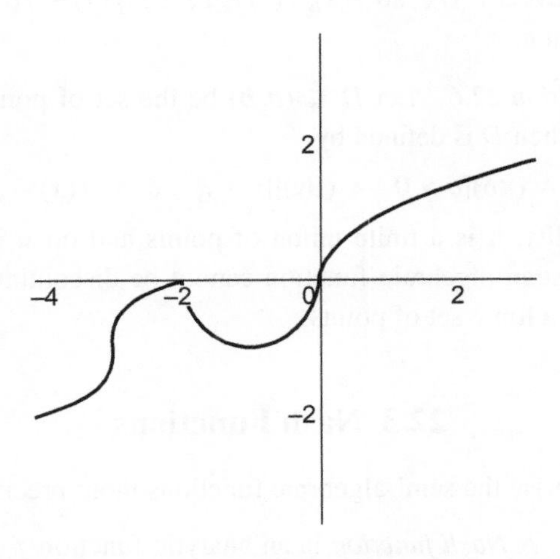

Figure 22.1 A function $\mathbb{R} \to \mathbb{R}$ which is definable in $\mathbb{R}_{\text{o-ring}}$ but which is not a polynomial function.

Proposition 22.6 *Let $(a, b) \subseteq \mathbb{R}$ be an open interval, possibly with $a = -\infty$ and / or $b = +\infty$, and let $f : (a, b) \to \mathbb{R}$ be a semi-algebraic function. Then f is* piecewise continuous, *that is, there is $m \in \mathbb{N}$ and there are $a_k \in \mathbb{R}$ with $a = a_0 < a_1 < \cdots < a_m < a_{m+1} = b$ such that for each $k = 0, \ldots, m$, the restriction $f{\restriction}_{(a_k,a_{k+1})}$ is continuous.*

There is more than one proof of this proposition. We give a proof which works for any o-minimal structure on the real line.

Lemma 22.7 *Let $f : (a, b) \to \mathbb{R}$ be a semi-algebraic function. Then there is $x \in (a, b)$ such that f is continuous at x.*

Proof First suppose that some point $c \in \mathbb{R}$ has infinite preimage under f. This preimage is defined by $f(x) = c$ so, by o-minimality, it contains an open interval U. Then $f{\restriction}_U$ is constant, so continuous.

Otherwise, we will construct a decreasing chain of intervals U_n. Set $U_0 = (a, b)$. Now assume $n \in \mathbb{N}^+$, and we have constructed U_{n-1}. Since no point in $\mathbb{R}$ has infinite preimage, the image of $f{\restriction}_{U_{n-1}}$ is infinite. This image is definable and so, by o-minimality, contains an open interval, I_n. We can choose this I_n to have length at most $1/n$. Then the set J_n defined by $x \in U_{n-1} \wedge f(x) \in I_n$ is infinite and so contains an open interval. We choose U_n to be some open interval in J_n such that the closure $\overline{U}_n$ is contained in U_{n-1}.

Then the intersection $\bigcap_{n\in\mathbb{N}} U_n = \bigcap_{n\in\mathbb{N}} \overline{U}_n$, an intersection of nested closed intervals, which is non-empty. Let $c \in \bigcap_{n\in\mathbb{N}} U_n$, and let $\epsilon > 0$. Take $N \in \mathbb{N}$ such that $N > 1/\epsilon$, and choose $\delta > 0$ such that $(c - \delta, c + \delta) \subseteq U_N$. Then, if $|x - c| < \delta$, we have $x \in U_N$, so $f(x), f(c) \in I_N$, so $|f(x) - f(c)| \leqslant 1/N < \epsilon$. So f is continuous at c. □

Proof of Proposition 22.6 Let $D \subseteq (a, b)$ be the set of points at which f is discontinuous. Then D is defined by

$$(\exists\epsilon)[\epsilon > 0 \wedge (\forall\delta)[\delta > 0 \to (\exists y)[|x - y| < \delta \wedge |f(x) - f(y)| > \epsilon]]],$$

so by o-minimality, it is a finite union of points and open intervals. But by Lemma 22.7, a semi-algebraic function cannot be discontinuous on an open interval. So D is a finite set of points. □

22.3 Nash Functions

We now characterise the semi-algebraic functions more precisely.

Definition 22.8 A *Nash function* is an analytic function $f : U \to \mathbb{R}$ where $U \subseteq \mathbb{R}^n$ is a connected open semi-algebraic subset such that there is a non-zero polynomial $h(\bar{x}, y) \in \mathbb{R}[\bar{x}, y]$ such that for all $\bar{x} \in U$, we have $h(\bar{x}, f(\bar{x})) = 0$.

To say that f is *analytic* means that it is infinitely differentiable and, furthermore, that for every point $a \in U$, the Taylor series of f at a converges to f in a neighbourhood of a. A polynomial function f is analytic and is a Nash function, taking $h(\bar{x}, y)$ to be $f(\bar{x}) - y$. The following consequence of the implicit function theorem for analytic functions gives an easy way to check analyticity.

Fact 22.9 *Suppose that $U \subseteq \mathbb{R}^n$ is a connected open semi-algebraic subset, that $f : U \to \mathbb{R}$ is continuous, and that there is a polynomial $h(\bar{x}, y) \in \mathbb{R}[\bar{x}, y]$ such that for all $\bar{x} \in U$, we have $h(\bar{x}, f(\bar{x})) = 0$ and $\frac{\partial h}{\partial y}(\bar{x}, f(\bar{x})) \neq 0$. Then f is a Nash function.*

Using this fact, the function f from Example 22.5 can be split into the three points $(-3, -1)$, $(-2, 0)$, and $(0, 0)$ and four Nash functions:

$$y = \begin{cases} \sqrt[3]{x+3} - 1 & \text{for } x \in (-\infty, -3) \text{ and for } x \in (-3, -2), \\ -\sqrt{1 - (x+1)^2} & \text{for } x \in (-2, 0), \\ \sqrt{x} & \text{for } x \in (0, \infty). \end{cases}$$

The functions $h(x, y)$ are then those appearing in the formula in Example 22.5. Note that f is not differentiable at $x = -3$, so is not analytic, and this corresponds to the function $h(x, y) = (y+1)^3 - x - 3$ having $\frac{\partial h}{\partial y}(-3, -1) = 0$.

Proposition 22.10 *Let $(a, b) \subseteq \mathbb{R}$ be an open interval, possibly with $a = -\infty$ and / or $b = +\infty$, and let $f : (a, b) \to \mathbb{R}$ be a semi-algebraic function. Then f is a* piecewise Nash function. *That is, there is $m \in \mathbb{N}$ and $a = a_0 < a_1 < \cdots < a_m < a_{m+1} = b$ such that, for each $k = 0, \ldots, m$, $f\restriction_{(a_k, a_{k+1})}$ is a Nash function.*

Proof By quantifier elimination for $\mathbb{R}_{\text{o-ring}}$ (Fact 18.5) and Lemma 22.2, the graph $\Gamma_f = \left\{(x, y) \in \mathbb{R}^2 \mid y = f(x)\right\}$ of f is defined by a formula of the form

$$\bigvee_{i=1}^{r} \left(h_i(x, y) = 0 \wedge \bigwedge_{j=1}^{s_i} g_{ij}(x, y) > 0 \right).$$

The functions h_i cannot be the zero polynomial, because otherwise Γ_f would contain an open subset of $\mathbb{R}^2$, which is impossible for the graph of a function.

Let H be the set consisting of the polynomials $h_i(x, y)$ and their iterated y-derivatives $\frac{\partial^t h_i}{\partial y^t}(x, y)$ for all $t \in \mathbb{N}$. Since the h_i are polynomials, H is a finite set. For each $h \in H$, let $Z_h = \{x \in \mathbb{R} \mid h(x, f(x)) = 0\}$ and let $D_h = \left\{x \in Z_h \,\middle|\, \frac{\partial h}{\partial y}(x, f(x)) \neq 0\right\}$. These sets are definable, so by o-minimality, they are finite unions of points and open intervals. List all of the isolated points and the endpoints of the intervals for all of the Z_h for $h \in H$ as $a = a_0 < a_1 < \cdots <$

$a_m < a_{m+1} = b$. Now we claim that each interval $I_k := (a_k, a_{k+1})$ is contained in one of the D_h. By definition, if $c, c' \in I_k$ and $h \in H$, then either both or neither of c and c' lie in Z_h. From the formula defining f, for some $i = 1, \dots, r$, we have $I_k \subseteq Z_{h_i}$. If $I_k \nsubseteq D_{h_i}$, then we replace h_i by $\frac{\partial h_i}{\partial y}$ and repeat if necessary. If d is the degree of $h_i(x, y)$ as a polynomial in y, then $\frac{\partial^d h_i}{\partial y^d}$ is constant in y and non-zero, so $\frac{\partial^d h_i}{\partial y^d}(x, f(x))$ is a non-zero polynomial in x which cannot vanish on the interval I_k. So at some stage we stop and find an $h \in H$ such that $I_k \subseteq D_h$.

Thus we have shown that f is a piecewise Nash function, as required. □

It is also true, and fairly easy to prove, that every piecewise Nash function $f : (a, b) \to \mathbb{R}$ is a semi-algebraic function, provided we use the definition given above that *piecewise* means split into only finitely many pieces.

22.4 Cells

The definable subsets of $\mathbb{R}^2$ can be built from Nash functions as follows.

Definition 22.11 A *0-cell* in $\mathbb{R}^2$ is a single point.

A *(Nash) 1-cell* is a subset of $\mathbb{R}^2$ defined by a formula of the form

$$x = c \wedge a < y \wedge y < b$$

or

$$a < x \wedge x < b \wedge y = f(x),$$

where $c \in \mathbb{R}$, $a < b$ with $a \in \{-\infty\} \cup \mathbb{R}$, and $b \in \mathbb{R} \cup \{+\infty\}$, and f is a Nash function on the interval (a, b). Conditions of the form $-\infty < x$ or $y < +\infty$ are automatically true and are not technically part of the L_{ring}-formula defining the cell. They are included above to streamline the definition of cells.

A *(Nash) 2-cell* is a subset of $\mathbb{R}^2$ defined by a formula of the form

$$a < x \wedge x < b \wedge f(x) < y \wedge y < g(x),$$

where $a < b$ are as above and f and g are Nash functions on (a, b) such that, for all $x \in (a, b)$, we have $f(x) < g(x)$, except that f may be the constant function with value $-\infty$ and g may be the constant function with value $+\infty$.

Examples of cells are given in Figure 22.2.

It is clear that Nash cells, and finite unions of them, are semi-algebraic. Using Proposition 22.10, one can prove the following.

Proposition 22.12 *Every semi-algebraic subset of $\mathbb{R}^2$ is a finite union of Nash cells.*

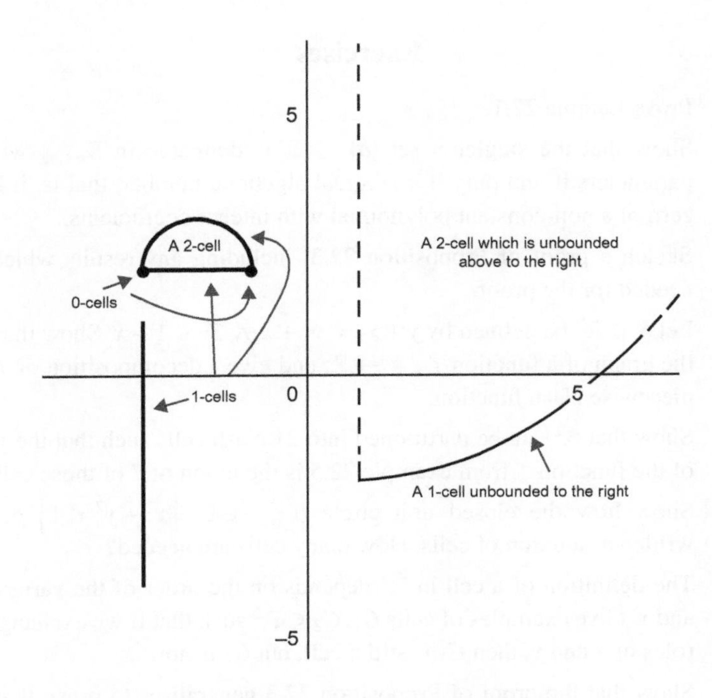

Figure 22.2 Some examples of cells in $\mathbb{R}^2$.

There are notions of Nash m-cells in $\mathbb{R}^n$ for all m and $n \in \mathbb{N}^+$, and the results analogous to Propositions 22.10 and 22.12 can be proved together by an induction on n.

If we replace Nash functions by arbitrary continuous definable functions in Definition 22.11, we get a more general notion of cells (continuous cells). Then the graph of the function f in Example 22.5 is a single 1-cell, whereas it is the union of seven Nash cells.

Remarkably, a generalisation of Proposition 22.12 to $\mathcal{M}^n$ for arbitrary $n \in \mathbb{N}^+$, called the *cell decomposition theorem*, is true in any o-minimal structure $\mathcal{M}$ when we consider continuous cells.

The study of o-minimal structures is an important branch of model theory. The book [vdD98] by van den Dries is an excellent introduction to the subject, and Chapter 2 treats the case of semi-algebraic sets in detail for the continuous cells. Semi-algebraic sets are studied in detail from a more geometric viewpoint in [BCR98]. More information about the model theory of $\mathbb{R}_{\text{o-ring}}$ can also be found in Section 3.3 of Marker's book [Mar02] and in the first chapter of [MMP06].

Exercises

22.1 Prove Lemma 22.1.

22.2 Show that the singleton set $\{a\} \subseteq \mathbb{R}$ is definable in $\mathbb{R}_{\text{o-ring}}$ without parameters if and only if a is a real algebraic number, that is, it is the zero of a non-constant polynomial with integer coefficients.

22.3 Sketch a proof of Proposition 22.3, including any results which are needed for the proof.

22.4 Let $S \subseteq \mathbb{R}^2$ be defined by $y + xy = y^2 + x \wedge 2y \leqslant 1 + x$. Show that S is the graph of a function $f : \mathbb{R} \to \mathbb{R}$, and give a decomposition of f as a piecewise Nash function.

22.5 Show that $\mathbb{R}^2$ can be partitioned into 21 Nash cells such that the graph of the function f from example 22.5 is the union of 7 of those cells.

22.6 Show how the closed unit circle $\left\{(x, y) \in \mathbb{R}^2 \,\middle|\, x^2 + y^2 \leqslant 1\right\}$ can be written as a union of cells. How many cells are needed?

22.7 The definition of a cell in $\mathbb{R}^2$ depends on the order of the variables x and y. Give examples of cells $C_1, C_2 \subseteq \mathbb{R}^2$ such that if we exchange the roles of x and y, then C_1 is still a cell, but C_2 is not.

22.8 Show that the proof of Proposition 22.3 generalises to prove that any real-closed field is o-minimal.

22.9 Use the completeness of RCF to show that the conclusions of Lemma 22.7 and Proposition 22.6 hold for any real-closed field.

22.10 Find a proof of Proposition 22.12.

22.11 [Uniform finiteness for semi-algebraic sets] Let $\varphi(\bar{x}, y)$ be an $L_{\text{o-ring}}$-formula. For each $\bar{a} \in \mathbb{R}^n$, write $D_{\bar{a}}$ for the subset of $\mathbb{R}$ defined by $\varphi(\bar{a}, y)$. Show that $D_{\bar{a}}$ is infinite if and only if it contains an open interval. Deduce that the set of $\bar{a}$ such that $D_{\bar{a}}$ is infinite is definable. Further deduce that there is $N \in \mathbb{N}$ such that for any $\bar{a} \in \mathbb{R}^n$, either $D_{\bar{a}}$ is infinite or $|D_{\bar{a}}| \leqslant N$.

22.12 Consider $\mathbb{R}_{\text{o-}\mathbb{R}\text{-VS}}$, the real line as an *ordered $\mathbb{R}$-vector space*, in the language $L_{\mathbb{R}\text{-VS}} \cup \{<\}$. Characterise the subsets of $\mathbb{R}^2$ which are parametrically definable by quantifier-free formulas in this structure. [This structure actually has quantifier elimination, and the definable sets are known as *semi-linear sets*. See [vdD98, pp25–28] for more details.]

22.13 Prove that $\left\{(x, y) \in \mathbb{R}^2 \,\middle|\, y = x^2\right\}$ is not definable in $\mathbb{R}_<$, nor even in $\mathbb{R}_{\text{o-}\mathbb{R}\text{-VS}}$.

Part V

Types

Types are a central concept in model theory. In this part of the book we start by explaining how types generalise elements or tuples from a structure: they are like potential elements which exist in elementary extensions. We then explain how types also generalise definable sets. A key tool is the omitting types theorem, which is proved by extending the method of Henkin that we previously used to prove the compactness theorem. The first application of types is the Ryll–Nardzewski theorem, which characterises the countably categorical structures as those structures $\mathcal{A}$ such that A^n has only finitely many definable subsets for each $n \in \mathbb{N}$. The ideas from the proof of that theorem naturally lead to an analysis of the countable models of other theories, including the notions of prime, atomic, universal, and $\aleph_0$-saturated models. Finally, we extend the idea of saturation to uncountable models and give a method for proving the completeness of a theory which, unlike the Łos-Vaught test, does not require the theory to be categorical in any cardinal. Here we need more set theory than elsewhere in the book, and some ideas are only sketched.

23
Realising Types

In Chapter 20 we saw that if (a_1, a_2) is a pair of elements from a $\mathbb{Q}$-vector space that are linearly dependent, then they satisfy a principal formula, but if they are linearly independent, then they do not. To analyse these tuples which do not satisfy a principal formula, we need to consider infinitely many formulas at once, which brings us to the notion of *types*.

23.1 Types

Recall that $\mathrm{Form}_n(L)$ is the set of all L-formulas with free variables from $x_1, \ldots, x_n$ only.

Definition 23.1 Let T be a complete L-theory. An *n-type* of T is a subset $p(\bar{x})$ of $\mathrm{Form}_n(L)$ such that $p(\bar{x})$ is *finitely satisfiable*, that is, for any formulas $\varphi_1(\bar{x}), \ldots, \varphi_r(\bar{x})$ from $p(\bar{x})$, $T \vdash \exists \bar{x} \, [\bigwedge_{i=1}^{r} \varphi_i(\bar{x})]$. The type $p(\bar{x})$ is a *complete type* if, for all $\varphi(\bar{x}) \in \mathrm{Form}_n(L)$, either $\varphi(\bar{x}) \in p(\bar{x})$ or $\neg\varphi(\bar{x}) \in p(\bar{x})$.

For example, we can consider the 2-type in the theory of infinite $\mathbb{Q}$-vector spaces consisting of all the formulas $\lambda_1 \cdot v_1 + \lambda_2 \cdot v_2 \neq 0$ such that $(\lambda_1, \lambda_2) \in \mathbb{Q}^2 \setminus \{(0,0)\}$. This is the type of a linearly independent pair of vectors.

Definition 23.2 Let T be an L-theory, let $\mathcal{M}$ be a model of T with domain M, and let $\bar{a} = (a_1, \ldots, a_n) \in M^n$. The *type of $\bar{a}$ in $\mathcal{M}$* is

$$\mathrm{tp}_{\mathcal{M}}(\bar{a}) = \{\varphi(\bar{x}) \in \mathrm{Form}_n(L) \mid \mathcal{M} \models \varphi(\bar{a})\}.$$

We should check that the type of a tuple really is a type, according to our definition.

Lemma 23.3 *Let $\mathcal{M} \models T$ and $\bar{a} \in M^n$. Then $\mathrm{tp}_{\mathcal{M}}(\bar{a})$ is a complete n-type of T.*

Proof The set of formulas $\text{tp}_{\mathcal{M}}(\bar{a})$ is finitely satisfiable, because it is satisfied by $\bar{a}$. For any $\varphi(\bar{x}) \in \text{Form}_n(L)$, either $\mathcal{M} \models \varphi(\bar{a})$ or $\mathcal{M} \models \neg\varphi(\bar{a})$, so $\text{tp}_{\mathcal{M}}(\bar{a})$ is complete. □

The type of a tuple is preserved under elementary extensions and automorphisms.

Proposition 23.4 (i) *Let* $\pi : \mathcal{M} \to \mathcal{N}$ *be an elementary embedding, and let* $\bar{a} \in M^n$. *Then* $\text{tp}_{\mathcal{M}}(\bar{a}) = \text{tp}_{\mathcal{N}}(\pi(\bar{a}))$.
(ii) *If* $\pi \in \text{Aut}(\mathcal{M})$ *and* $\bar{a} \in M^n$, *then* $\text{tp}_{\mathcal{M}}(\bar{a}) = \text{tp}_{\mathcal{M}}(\pi(\bar{a}))$.

Proof For (i), we have $\varphi(\bar{x}) \in \text{tp}_{\mathcal{M}}(\bar{a})$ iff $\mathcal{M} \models \varphi(\bar{a})$ iff $\mathcal{N} \models \varphi(\pi(\bar{a}))$ by definition of an elementary embedding, and this is iff $\varphi(\bar{x}) \in \text{tp}_{\mathcal{N}}(\pi(\bar{a}))$. Now note that (ii) is a special case of (i). □

Usually we are only considering one model, or elementary extensions or elementary submodels of a given model, so then, by the proposition, the type of tuple does not depend on the particular model $\mathcal{M}$. So usually we write just $\text{tp}(\bar{a})$ without the subscript $\mathcal{M}$.

23.2 Realising Types

Definition 23.5 Let T be a complete theory, $\mathcal{M}$ a model of T, and $p(\bar{x})$ an n-type of T. We say that the type p is *realised in* $\mathcal{M}$ iff there is a tuple $\bar{a} \in M^n$ such that, for all $\varphi(\bar{x}) \in p(\bar{x})$, $\mathcal{M} \models \varphi(\bar{a})$. In this case we write $\mathcal{M} \models p(\bar{a})$. If p is not realised in $\mathcal{M}$, we say that it is *omitted from* $\mathcal{M}$.

For example, the type of a $\mathbb{Q}$-linearly independent pair of vectors is realised in a $\mathbb{Q}$-vector space of dimension at least 2 but omitted in a model of dimension 1. Using the compactness theorem and the method of new constants, we can show that every type of T can be realised in some model of T.

Lemma 23.6 *Suppose that p is an n-type of T, a complete L-theory. Then there is a model $\mathcal{M}$ of T of cardinality at most $|L|$ and $\bar{a} \in M^n$ such that $\mathcal{M} \models p(\bar{a})$.*

Proof Expand L to L' by adding new constant symbols $c_1, \ldots, c_n$. Consider the set of sentences $\Sigma = T \cup \{\varphi(\bar{c}) \mid \varphi(\bar{x}) \in p(\bar{x})\}$. Then Σ is finitely satisfiable since $p(\bar{x})$ is a type, so by the strong version of the compactness theorem, Theorem 11.7, it has a model $\mathcal{M}'$ of cardinality at most $|L'|$, which is equal to $|L|$. Take $\bar{a} = \bar{c}^{\mathcal{M}'}$, and let $\mathcal{M}$ be the reduct to L. Then $\mathcal{M} \models p(\bar{a})$. □

In fact, we can realise as many types as we like in the same model.

Proposition 23.7 *Let P be any set of types of a complete theory T, and let $\mathcal{M} \models T$. Then there is an elementary extension of $\mathcal{M}$ which realises all the types in P.*

We leave the proof as an exercise for the reader.

23.3 Some Types in Th($\mathbb{N}_{\text{s-ring}}$)

We will investigate some 1-types in the structure $\mathbb{N}_{\text{s-ring}}$. Firstly, for each $n \in \mathbb{N}$, we can consider tp(n). If $n \neq m$, then tp(n) $\neq$ tp(m), because the formula $\psi_n(x)$ given by $x = 0 \underbrace{+ 1 + \cdots + 1}_{n}$ is in tp(n) but not in tp(m). These are all the types which are realised in $\mathbb{N}_{\text{s-ring}}$ itself, but there are many more types which are not. Let $p_{ns}(x) = \{\neg\psi_n(x) \mid n \in \mathbb{N}\}$. Then p_{ns} is finitely satisfiable and so is a type, so it is realised in some model of Th($\mathbb{N}_{\text{s-ring}}$) other than the standard model $\mathbb{N}_{\text{s-ring}}$. Any such model $\mathcal{N}$ is called a *non-standard model of arithmetic*, and an element of $\mathcal{N}$ which realises p_{ns} is called a *non-standard natural number*.

Let $\theta(x)$ be the formula $\exists y[y + y = x]$ which states that x is even. Then $p_{ns}(x) \cup \{\theta(x)\}$ and $p_{ns}(x) \cup \{\neg\theta(x)\}$ are each finitely satisfiable, so the type of a non-standard natural number splits into the types of even and odd non-standard numbers. In particular, p_{ns} is not a complete type. (The more meaningful observation here is that the deductive closure of p_{ns} is not a complete type. However, it is useful to identify types with their deductive closures, just as we often identify a list of axioms with the theory it axiomatises.)

In fact, if $S \subseteq \mathbb{N}$ is any infinite definable subset, say, defined by the formula $\varphi(x)$, then $p(x) \cup \{\varphi(x)\}$ is finitely satisfiable and hence a type. Thus we can talk about 'non-standard members of S'.

The formula $\rho_{ns}(x)$ given by $\forall yz[x = y \cdot z \to (x = y \vee x = z)] \wedge x \neq 0 \wedge x \neq 1$ defines the set of all prime numbers in $\mathbb{N}$. Since there are infinitely many primes in $\mathbb{N}$, there are also non-standard primes. By definition, a non-standard prime is not divisible by any standard natural number. The opposite behaviour is also possible.

Lemma 23.8 *There is a 1-type of* Th($\mathbb{N}_{\text{s-ring}}$) *of a non-standard natural number which is divisible by every standard prime number.*

Proof The formula $\delta(x, y)$ given by $\exists z[y \cdot z = x]$ states that y divides x. Take $p(x)$ to be the set of formulas $\{\delta(x, P) \mid P \text{ is a standard prime number}\}$. If $\delta(x, P_1), \ldots, \delta(x, P_r)$ are finitely many formulas from $p(x)$, then the number $N = \prod_{i=1}^{r} P_i$ satisfies them all. So $p(x)$ is finitely satisfiable, hence it is a type. □

Exercises

23.1 Suppose $\mathcal{A}$ is a substructure of $\mathcal{B}$. Show that $\mathcal{A} \preccurlyeq \mathcal{B}$ if and only if, for every finite tuple $\bar{a}$ from $\mathcal{A}$, $\mathrm{tp}_{\mathcal{A}}(\bar{a}) = \mathrm{tp}_{\mathcal{B}}(\bar{a})$.

23.2 Prove Proposition 23.7.

23.3 Show that there is a 1-type of $\mathrm{Th}(\mathbb{N}_{\text{s-ring}})$ of a number whose only prime factor is 5 but which is divisible by 5^n for all $n \in \mathbb{N}$. Show there is another 1-type of $\mathrm{Th}(\mathbb{N}_{\text{s-ring}})$ of a non-standard number which is divisible by 3 but not by 9 or by any other standard natural number greater than 3.

23.4 Suppose $\bar{a}, \bar{b} \in \mathbb{Q}^n$ with $a_1 < a_2 < \cdots < a_n$ and $b_1 < b_2 < \cdots < b_n$. By considering automorphisms of $\mathbb{Q}_<$, show that $\mathrm{tp}_{\mathbb{Q}_<}(\bar{a}) = \mathrm{tp}_{\mathbb{Q}_<}(\bar{b})$. Deduce that there are only finitely many complete n-types in DLO, for each $n \in \mathbb{N}^+$.

23.5 Show there are exactly two complete 1-types in $T^{\infty}_{K\text{-VS}}$ for any field K.

23.6 Let $p(\bar{x})$ be an n-type of a complete L-theory T. Let $L' = L \cup \{c_1, \ldots, c_n\}$, with the c_i being new constant symbols, and let T' be the deductive closure of $\{\varphi(\bar{c}) \mid \varphi(\bar{x}) \in p(\bar{x})\}$. Show that T' is an L'-theory and that p is a complete type if and only if T' is a complete L'-theory.

23.7 Recall that $\mathrm{RCF} = \mathrm{Th}(\mathbb{R}_{\text{o-ring}})$, and let $p_{+\infty}(x) = \{x > n \mid n \in \mathbb{N}\}$. Show that $p_{+\infty}$ is a 1-type of RCF. For every polynomial $f(x) \in \mathbb{Z}[x]$, show that exactly one of the formulas $f(x) = 0$, $f(x) < 0$, or $f(x) > 0$ is in (the deductive closure of) $p_{+\infty}$. Using quantifier elimination for RCF, deduce that $p_{+\infty}$ is a complete type of RCF.

23.8 Recall from Exercise 18.4 that $T_S = \mathrm{Th}(\mathbb{N}_{\text{succ}})$ has quantifier elimination. Use this fact to determine all of the complete n-types in T_S for $n \in \mathbb{N}^+$.

24
Omitting Types

In the previous chapter we saw that any type of a theory T can be realised in some model of T. In this chapter we consider which types are realised in all models of T and which types can be omitted. First we explain how types can be considered as a generalisation of definable sets.

24.1 Types as Intersections of Definable Sets

Given a complete theory T and a model $\mathcal{M} \models T$, a formula $\varphi(\bar{x})$ defines the subset $\varphi(\mathcal{M}) = \{\bar{a} \in M^n \mid \mathcal{M} \models \varphi(\bar{a})\}$. Likewise, given an n-type $p(\bar{x})$ of T, we can write $p(\mathcal{M}) = \{\bar{a} \in M^n \mid \text{for all } \varphi(\bar{x}) \in p(\bar{x}), \mathcal{M} \models \varphi(\bar{a})\}$. So $p(\mathcal{M})$ is the intersection $\bigcap \{\varphi(\mathcal{M}) \mid \varphi(\bar{x}) \in p(\bar{x})\}$.

For example, if V is a $\mathbb{Q}$-vector space, we can consider the definable sets

$$S_{\lambda_1,\lambda_2} = \left\{(v_1, v_2) \in V^2 \mid \lambda_1 \cdot v_1 + \lambda_2 \cdot v_2 \neq 0\right\}$$

for each $(\lambda_1, \lambda_2) \in \mathbb{Q}^2 \smallsetminus \{(0,0)\}$. Then the intersection of the sets S_{λ_1,λ_2} is the set of linearly independent pairs of vectors in the model V.

More generally, we can consider the intersection of any set $\{\varphi_i(\bar{x}) \mid i \in I\}$ of definable subsets of M^n. The condition that the $\varphi_i(\bar{x})$ are *finitely satisfiable*, and so are actually a type, corresponds to the condition that any intersection of finitely many of the sets $\varphi_i(\mathcal{M})$ is non-empty.

The intersection $p(\mathcal{M})$ may be empty, and this corresponds to the type p being omitted by the model $\mathcal{M}$. For example, in the case of vector spaces above, the intersection $\bigcap \left\{S_{\lambda_1,\lambda_2} \mid (\lambda_1, \lambda_2) \in \mathbb{Q}^2 \smallsetminus \{(0,0)\}\right\}$ is empty if V is a one-dimensional $\mathbb{Q}$-vector space but non-empty if $\dim V \geqslant 2$.

So we cannot necessary refer to the set of realisations in a particular model as the type. However, by Proposition 23.7, there is a model $\mathcal{M}$ of T which realises all the n-types of T, for all $n \in \mathbb{N}^+$. In this case, if $p(\bar{x})$ and $q(\bar{x})$ are different n-types, then $p(\mathcal{M})$ and $q(\mathcal{M})$ are different subsets of $\mathcal{M}^n$. If $\mathcal{M}$ is such a model, then we refer to sets of the form $p(\mathcal{M})$ as *type-definable subsets* of M^n.

Remarks 24.1 (i) Type-definable sets are a generalisation of definable sets, and they show how we can think of types as a generalisation of formulas. On the other hand, complete types are like elements (or tuples of elements) from a model, or potential elements, and so types are somehow a common abstraction of elements and of definable sets. This idea is important in the branch of model theory known as *stability theory*.
(ii) Sometimes what we have called *types* are known as *partial types*, and the word *type* is used to mean *complete type*.

24.2 Omitting Types

We now consider which types can be omitted. Roughly speaking, if a type can be given by a single formula, then it must be realised in any model of T. Otherwise, it can be omitted in some models. We can make this more precise.

Definition 24.2 An n-type $p(\bar{x})$ of a complete theory T is a *principal type* if there is a single formula $\psi(\bar{x})$ such that $T \vdash \exists\bar{x}\psi(\bar{x})$ and, for every formula $\varphi(\bar{x}) \in p(\bar{x})$, we have $T \vdash \forall\bar{x}[\psi(\bar{x}) \to \varphi(\bar{x})]$. Any type which is not principal is called a *non-principal type*.

It is easy to see that any principal type of T is realised in any model of T. What is surprising is that the converse is also true (at least when the language is countable).

Theorem 24.3 (Omitting types theorem) *Let T be a complete theory in a countable language L, and let p be a non-principal n-type of T. Then there is a countable model of T which omits p.*

We will adapt the proof of Proposition 11.4 which shows that any finitely satisfiable set of L-sentences can be extended to a Henkin theory T^* in a language with extra constant symbols. By Proposition 11.6, every Henkin theory has a canonical model, that is, a model in which every element is named by a closed term. We will construct our Henkin theory T^* in such a way that every element of the canonical model $\mathcal{M}^*$ is named by a new constant symbol, and for every n-tuple $\bar{c}$ of these constant symbols, there is a formula $\varphi(\bar{x}) \in p(\bar{x})$

such that $T^* \vdash \neg\varphi(\bar{c})$. It follows that no n-tuple from $\mathcal{M}^*$ realises p, so the reduct $\mathcal{M}$ of $\mathcal{M}^*$ to L is a model of T which omits p.

Proof of Theorem 24.3 Let L^* be L expanded by a countably infinite set C of new constant symbols. Enumerate all the n-tuples of these new constant symbols as $(\bar{c}_r)_{r\in\mathbb{N}}$, and enumerate all the L^*-sentences as $(\sigma_r)_{r\in\mathbb{N}}$. We will construct L^*-sentences θ_m for $m \in \mathbb{N}$ such that for each $m \in \mathbb{N}$ we have $\theta_{m+1} \vdash \theta_m$, $T \cup \{\theta_m\}$ is satisfiable and:

(i) if $m = 3k + 1$, either $\theta_m \vdash \sigma_k$ or $\theta_m \vdash \neg\sigma_k$;
(ii) if $m = 3k + 2$, $\theta_{m-1} \vdash \sigma_k$ and σ_k is of the form $\exists x[\varphi(x)]$, then $\theta_m \vdash \varphi(c)$ for some $c \in C$; and
(iii) if $m = 3k + 3$, there is some formula $\varphi(\bar{x}) \in p(\bar{x})$ such that $\theta_m \vdash \neg\varphi(\bar{c}_k)$.

To do this, we proceed by induction on m. Take θ_0 to be $\forall x[x = x]$. Then certainly $T \cup \{\theta_0\}$ is satisfiable. Now suppose $m \in \mathbb{N}^+$ and we have constructed $\theta_0, \ldots, \theta_{m-1}$.

If m is of the form $3k + 1$ for some $k \in \mathbb{N}$, take $\theta_m = \theta_{m-1} \wedge \sigma_k$, if $T \cup \{\theta_{m-1} \wedge \sigma_k\}$ is satisfiable, and take $\theta_m = \theta_{m-1} \wedge \neg\sigma_k$ otherwise. Then $T \cup \{\theta_m\}$ is satisfiable.

If m is of the form $3k+2$, $\theta_{m-1} \vdash \sigma_k$ and σ_k is of the form $\exists x[\varphi(x)]$, choose a constant symbol $c \in C$ which does not appear in θ_{m-1} and set $\theta_m = \theta_{m-1} \wedge \varphi(c)$. Then $T \cup \{\theta_m\}$ is satisfiable because c is a new constant. Otherwise, we just let $\theta_m = \theta_{m-1}$.

If m is of the form $3k + 3$, we claim there is a formula $\varphi(\bar{x}) \in p(\bar{x})$ such that $T \cup \{\theta_{m-1}, \neg\varphi(\bar{c}_k)\}$ is satisfiable. Then we can take $\theta_m = \theta_{m-1} \wedge \neg\varphi(\bar{c}_k)$. To prove the claim, we write θ_{m-1} in the form $\rho(\bar{c}_k, \bar{d})$, where $\rho(\bar{x}, \bar{y})$ is an L-formula and $\bar{d}$ is a tuple of constants from C, all distinct from the elements of $\bar{c}_k$. Then, if there is no $\varphi(\bar{x}) \in p(\bar{x})$ such that $T \cup \{\theta_{m-1}, \neg\varphi(\bar{c}_k)\}$ is satisfiable, for each $\varphi(\bar{x}) \in p(\bar{x})$ we have $T \vdash \rho(\bar{c}_k, \bar{d}) \to \varphi(\bar{c}_k)$, but none of the constants in $\bar{c}_k$ or in $\bar{d}$ occur in T, so we must have $T \vdash \forall\bar{x}\forall\bar{y}[\rho(\bar{x}, \bar{y}) \to \varphi(\bar{x})]$, or equivalently, $T \vdash \forall\bar{x}[\exists\bar{y}[\rho(\bar{x}, \bar{y})] \to \varphi(\bar{x})]$. But then $\exists\bar{y}[\rho(\bar{x}, \bar{y})]$ is a principal formula for $p(\bar{x})$, a contradiction.

Thus we can produce our list of L^*-sentences θ_m for $m \in \mathbb{N}$ satisfying all the conditions.

Now take T^* to be the deductive closure of $T \cup \{\theta_m \mid m \in \mathbb{N}\}$. Then T^* is finitely satisfiable, it is complete by condition (i), and it has the witness property by condition (ii), and so it is a Henkin theory. Using Proposition 11.6, let $\mathcal{M}^*$ be a canonical model of T^* and $\mathcal{M}$ its reduct to L. By condition (ii), every element of M is named by a constant symbol from C, and so $\mathcal{M}$ is countable. By condition (iii), no n-tuple of elements of M realises p. So $\mathcal{M}$ omits p, as required. □

The theorem can be improved. More or less the same proof, with a little more organisation, gives the following.

Theorem 24.4 (Vaught's omitting types theorem) *Let T be a complete theory in a countable language L, and let P be a countable set of non-principal types of T. Then there is a countable model of T which omits all the types in P.*

Exercises

24.1 Let $p(\bar{x})$ be an n-type of $\text{Th}(\mathcal{M})$. Show that p is principal if and only if there is a non-empty definable set $S \subseteq M^n$ such that $S \subseteq p(\mathcal{M})$.

24.2 In $\text{Th}(\mathbb{N}_{\text{s-ring}})$, give an example of a complete principal 1-type and an incomplete non-principal 1-type.

24.3 Show that if $p(\bar{x})$ is a complete principal type, then there is a principal formula $\psi(\bar{x})$ in $p(\bar{x})$. Find an example of an incomplete principal type which does not contain a principal formula.

24.4 In $\mathbb{R}_{\text{o-ring}}$, show that the family of intervals $\{(0, 1 + 1/n) \mid n \in \mathbb{N}^+\}$ is a principal 1-type and the family of intervals $\{(1, 1 + 1/n) \mid n \in \mathbb{N}^+\}$ is a non-principal 1-type.

24.5 Let $r \in \mathbb{R}$ be a *real algebraic number*, that is, a root of a non-zero polynomial $f(x) \in Z[x]$. Show that $\text{tp}_{\mathbb{R}_{\text{o-ring}}}(r)$ is a principal type. Using quantifier elimination for $\mathbb{R}_{\text{o-ring}}$, show that if $r \in \mathbb{R}$ is a transcendental number, then $\text{tp}_{\mathbb{R}_{\text{o-ring}}}(r)$ is non-principal.

24.6 Let K be a field and let $V \models T^{\infty}_{K\text{-VS}}$. Suppose $(a_1, \ldots, a_n) \in V^n$ is linearly independent and $(b_1, \ldots, b_n) \in V^n$ is also linearly independent.

(a) Show that $\text{tp}(\bar{a}) = \text{tp}(\bar{b})$.
(b) Suppose K is a finite field. Show $\text{tp}(\bar{a})$ is a principal type.
(c) Suppose K is an infinite field and $n \geqslant 2$. Show $\text{tp}(\bar{a})$ is not principal.

24.7 Here is an example of a theory with no principal complete types. Let B be the set of all binary sequences $b = (b_n)_{n \in \mathbb{N}^+}$, with each $b_n \in \{0, 1\}$. We make B into an L-structure $\mathcal{B}$ by taking L to be the language with unary relation symbols Z_d for each $d \in \mathbb{N}^+$, where $Z_d^{\mathcal{B}} = \{b \in B \mid b_d = 0\}$. Show that $\text{Th}(\mathcal{B})$ has no principal complete 1-types.

24.8 Sketch a proof of the omitting types theorem, including any lemmas which are needed for its proof. Give all the key ideas and explain how they fit together, without giving all the details.

24.9 Prove Theorem 24.4.

25
Countable Categoricity

In Chapter 15 we saw that the theory DLO of dense linear orders without endpoints is countably categorical, that is, up to isomorphism, it has exactly one countably infinite model, which is $\mathbb{Q}_<$. Then, in Chapter 17, we saw that for each $n \in \mathbb{N}^+$, only finitely many subsets of $\mathbb{Q}^n$ are definable in $\mathbb{Q}_<$. In the terminology of Chapter 20, the Lindenbaum algebras $\text{Lind}_n(\text{DLO})$ are all finite.

In this chapter we show that this finiteness property characterises countably categorical theories. The proof uses types, and there are two more equivalent properties involving types.

25.1 Stone Spaces

First we note that we do not lose much by considering only complete types.

Lemma 25.1 *Let T be a complete theory. Then every n-type of T is contained in a complete n-type of T.*

Proof Let $p(\bar{x})$ be an n-type of T. By Lemma 23.6, there is a model $\mathcal{M}$ of T and $\bar{a} \in M^n$ such that $\mathcal{M} \models p(\bar{a})$. Then $\text{tp}_{\mathcal{M}}(\bar{a}) \supseteq p(\bar{x})$, and $\text{tp}_{\mathcal{M}}(\bar{a})$ is a complete n-type of T. □

Definition 25.2 (Stone space) For a complete theory T, let $S_n(T)$ be the set of all complete n-types of T. It is called the nth *Stone space* of T.

Proposition 25.3 *Let T be a complete theory and $n \in \mathbb{N}^+$. The following are equivalent:*

(i) *Every type in $S_n(T)$ is principal.*
(ii) *The Stone space $S_n(T)$ is finite.*
(iii) *The Lindenbaum algebra* $\text{Lind}_n(T)$ *is finite.*

Proof (i) $\Rightarrow$ (ii): Suppose every type in $S_n(T)$ is principal, but suppose for a contradiction that $S_n(T)$ is infinite, say $S_n(T) = \{p_i \mid i \in I\}$. Let ψ_i be a principal formula for p_i. Let $q = \{\neg\psi_i \mid i \in I\}$. We claim that q is finitely satisfiable. So let $\neg\psi_{i_1}, \dots, \neg\psi_{i_r}$ be a finite subset of q. Since I is infinite, there is $i \in I \setminus \{i_1, \dots, i_r\}$. Let $\bar{a}$ be an n-tuple in some model $\mathcal{M}$ such that $\text{tp}(\bar{a}) = p_i$. Then $\mathcal{M} \models \bigwedge_{j=1}^{r} \neg\psi_{i_j}(\bar{a})$. So q is finitely satisfiable and hence is a type. By Lemma 25.1, there is a complete type q' containing q. But q' cannot be any of the p_i for $i \in I$, which is a contradiction. So $S_n(T)$ is finite.

(ii) $\Rightarrow$ (i): Suppose $S_n(T)$ is finite, say, $S_n(T) = \{p_1, \dots, p_r\}$. For each $i, j \in \{1, \dots, r\}$ with $i \neq j$, there is a formula $\varphi_{ij}(\bar{x})$ such that $\varphi_{ij}(\bar{x}) \in p_i$ and $\neg\varphi_{ij}(\bar{x}) \in p_j$. Let $\psi_i = \bigwedge_{j=1, j\neq i}^{r} \varphi_{ij}$. Then $T \vdash \exists\bar{x}\psi_i(\bar{x})$ because p_i is a type and so finitely satisfiable. Now suppose $\mathcal{M} \models T$ and $\bar{a} \in M^n$ with $\mathcal{M} \models \psi_i(\bar{a})$. Then, for each $j \neq i$, $\mathcal{M} \models \varphi_{ij}(\bar{a})$, so $\text{tp}_{\mathcal{M}}(\bar{a}) \neq p_j$. Thus $\text{tp}_{\mathcal{M}}(\bar{a}) = p_i$. In particular, ψ_i is a principal formula for p_i. So every $p \in S_n(T)$ is principal.

(iii) $\Rightarrow$ (ii): Each type in $S_n(T)$ corresponds to a subset of $\text{Lind}_n(T)$, and different types correspond to different subsets. There are $2^{|\text{Lind}_n(T)|}$ subsets of $\text{Lind}_n(T)$, so $|S_n(T)| \leqslant 2^{|\text{Lind}_n(T)|}$. Thus, if $\text{Lind}_n(T)$ is finite, so is $S_n(T)$.

(ii) $\Rightarrow$ (iii): Suppose $S_n(T)$ is finite. Let $r = |S_n(T)|$. Then, by (ii) $\Rightarrow$ (i), every type in $S_n(T)$ is principal. Say the principal formulas are $\psi_1(\bar{x}), \dots, \psi_r(\bar{x})$. Then, for each subset J of $\{1, \dots, r\}$, the formula $\varphi_J(\bar{x}) = \bigvee_{j\in J} \psi_j(\bar{x})$ defines a different subset of M^n. But if S is a definable subset of M^n, and p is a type, then either every realisation of p is in S or no realisation of p is in S, so S must be defined by one of the formulas $\varphi_J(\bar{x})$. Hence $|\text{Lind}_n(T)| = 2^{|S_n(T)|}$, so $|\text{Lind}_n(T)|$ is finite. □

25.2 The Ryll–Nardzewski Theorem

Now we can state and prove the main theorem characterising countably categorical theories. We use the results on realising and omitting types and also the back-and-forth method.

Theorem 25.4 (Ryll–Nardzewski, Svenonius, Engeler) *Let T be a complete theory with infinite models in a countable language L. The following are equivalent.*

(i) *T is countably categorical.*
(ii) *For all $n \in \mathbb{N}^+$, every type in $S_n(T)$ is principal.*
(iii) *For all $n \in \mathbb{N}^+$, the Stone space $S_n(T)$ is finite.*
(iv) *For all $n \in \mathbb{N}^+$, the Lindenbaum algebra $\text{Lind}_n(T)$ is finite.*

Proof Conditions (ii), (iii), and (iv) are all equivalent by Proposition 25.3.

(i) $\Rightarrow$ (ii): Suppose that T is countably categorical but, for a contradiction, assume there is a non-principal type $p \in S_n(T)$ for some $n \in \mathbb{N}^+$. By the omitting types theorem, Theorem 24.3, there is a countable model $\mathcal{A}$ of T which omits p. By Lemma 23.6, there is a countable model $\mathcal{B}$ of T and $\bar{a} \in B^n$ such that $\text{tp}(\bar{a}) = p$. Since T is countably categorical, there is an isomorphism $\pi : \mathcal{B} \to \mathcal{A}$. Then, by Proposition 23.4, we have $\text{tp}_{\mathcal{A}}(\pi(\bar{a})) = \text{tp}_{\mathcal{B}}(\bar{a}) = p$. But $\mathcal{A}$ omits p, so we have a contradiction. So every complete type of T is principal.

(ii) $\Rightarrow$ (i): This part of the proof uses the back-and-forth method and is very similar to the proof that DLO is $\aleph_0$-categorical. Firstly, since the language is countable and T has infinite models, by the strong compactness theorem, there is at least one model of cardinality $\aleph_0$.

Suppose $\mathcal{A}$ and $\mathcal{B}$ are both models of T of cardinality $\aleph_0$. Enumerate A as $(a_n)_{n\in\mathbb{N}^+}$ and B as $(b_n)_{n\in\mathbb{N}^+}$. We will construct new enumerations $(\alpha_n)_{n\in\mathbb{N}^+}$ of A and $(\beta_n)_{n\in\mathbb{N}^+}$ of B such that for every $n \in \mathbb{N}$ and every L-formula $\varphi(x_1, \ldots, x_n)$, we have

$$\mathcal{A} \models \varphi(\alpha_1, \ldots, \alpha_n) \quad \text{if and only if} \quad \mathcal{B} \models \varphi(\beta_1, \ldots, \beta_n). \tag{$*$}$$

We proceed by induction on n. For $n = 0$, condition $(*)$ says that every L-sentence is true in $\mathcal{A}$ iff it is true in $\mathcal{B}$. This is correct, because $\mathcal{A}$ and $\mathcal{B}$ are models of the same complete L-theory.

Now suppose we have α_i and β_i for all $i < n$. If $n = 2m - 1$, odd, then let $\alpha_n = a_m$. Let $\psi_n(x_1, \ldots, x_n)$ be a principal formula for $(\alpha_1, \ldots, \alpha_n)$. Then we have $\mathcal{A} \models \psi_n(\alpha_1, \ldots, \alpha_n)$, and so we get $\mathcal{A} \models \exists x_n[\psi_n(\alpha_1, \ldots, \alpha_{n-1}, x_n)]$. Then by induction using $(*)$, it follows that $\mathcal{B} \models \exists x_n[\psi_n(\beta_1, \ldots, \beta_{n-1}, x_n)]$. Choose $\beta_n \in B$ such that $\mathcal{B} \models \psi_n(\beta_1, \ldots, \beta_n)$. Then, since ψ_n is a principal formula, condition $(*)$ holds for n.

If $n = 2m$, even, then let $\beta_n = b_m$. Repeat the above argument, swapping the roles of $\mathcal{A}$ and $\mathcal{B}$, to choose α_n such that condition $(*)$ holds for n.

Note that every a_m appears as some α_n. It may appear more than once, but if $\alpha_n = \alpha_l$, then this is witnessed by a formula, so also $\beta_n = \beta_l$ by $(*)$. Thus we may define a function $\pi : A \to B$ by $\pi(\alpha_n) = \beta_n$. This function is surjective, since every b_m appears as some β_n. It is injective, since if $\beta_n = \beta_l$, then also $\alpha_n = \alpha_l$ by $(*)$. Now $(*)$ also tells us that π preserves all formulas, in particular it preserves all atomic formulas and their negations, so it is an embedding. The same argument shows that π^{-1} is an embedding, so π is an isomorphism. So $\mathcal{A} \cong \mathcal{B}$, as required. Thus T is countably categorical. □

Exercises

25.1 Let T be any complete theory. Show that a formula $\psi(\bar{x})$ is a principal formula for T if and only if $[\psi(\bar{x})]$ is an atom of the Boolean algebra $\mathrm{Lind}_n(T)$.

25.2 Use the idea from the proof of Lemma 20.6 to show that if $\mathrm{Lind}_n(T)$ is finite, then every type in $S_n(T)$ is principal.

25.3 Suppose that every complete n-type of a complete theory T is principal. Show that every incomplete n-type of T is also principal.

25.4 Show that the theories $\mathrm{ACF}_0 = \mathrm{Th}(\mathbb{C}_{\mathrm{ring}})$ and $\mathrm{RCF} = \mathrm{Th}(\mathbb{R}_{\text{o-ring}})$ are not $\aleph_0$-categorical.

25.5 List all of the complete 2-types in the theory $T^{\infty}_{\mathbb{F}_3\text{-VS}}$. How many complete 3-types are there?

25.6 Sketch a proof of the Ryll–Nardzewski theorem. Explain all the key ideas in the proof and how they fit together, without writing out all the details.

25.7 Suppose that $T = \mathrm{Th}(\mathcal{A})$, that $|A| = \aleph_0$, and that every finite tuple $\bar{a}$ from A realises a principal type. By adapting the last part of the proof of the Ryll-Nardzewski theorem, show that for any model $\mathcal{B} \models T$, there is an elementary embedding $\pi : \mathcal{A} \to \mathcal{B}$.

25.8 Countably categorical theories can also be characterised in terms of permutation groups. This exercise and the next one briefly explain the idea. Suppose $\mathcal{A}$ is a countable L-structure whose theory is countably categorical. Let $\bar{a}, \bar{b} \in A^n$ have the same type. Use the back-and-forth method to construct $\pi \in \mathrm{Aut}(\mathcal{A})$ such that $\pi(\bar{a}) = \bar{b}$.

25.9 Using the Ryll–Nardzewski theorem, we can deduce that the permutation action of $\mathrm{Aut}(\mathcal{A})$ on A^n has only finitely many orbits, for each $n \in \mathbb{N}^+$. A group G acting on an infinite set A such that the induced action on A^n has only finitely many orbits is called an *oligomorphic permutation group*.

Let G be an oligomorphic permutation group on a countably infinite set A. Name each G-orbit O of A^n by an n-ary relation symbol R_O, to make a structure $\mathcal{A}$ in a language L. Show that every quantifier-free definable subset of A^n is a finite union of orbits and that the projection of such a set to A^{n-1} is still quantifier-free definable, so $\mathcal{A}$ has quantifier elimination. Show that $\mathcal{A}$ is countably categorical.

26
Large and Small Countable Models

In the previous chapter we considered theories with exactly one countable model. There are other theories, such as $T^{\infty}_{\mathbb{Q}\text{-VS}}$, where we can make sense of the idea that some countable models are smaller than others. In cardinality they are the same size, but the vector space $\mathbb{Q}^1$ of dimension 1 is in a sense smaller than the the space $\mathbb{Q}^3$ of dimension 3, which in turn is smaller than the vector space $\mathbb{Q}^{\oplus\aleph_0}$ of dimension $\aleph_0$. One way to see this is that there are elementary embeddings $\mathbb{Q}^1 \preccurlyeq \mathbb{Q}^3 \preccurlyeq \mathbb{Q}^{\oplus\aleph_0}$, but not the other way around. In this chapter we consider notions of the smallest and largest countable models and give a sufficient condition on a theory T for them to exist. Most of the proofs involve constructing an elementary embedding inductively, one element at a time, sometimes using the back-and-forth method. These proofs are very similar to each other and to proofs we have done earlier, so they are left as exercises.

26.1 Atomic and Prime Models

Let T be a complete theory with infinite models.

Definition 26.1 A model $\mathcal{A} \models T$ is *prime* if, for every model $\mathcal{M} \models T$, there is an elementary embedding $\mathcal{A} \preccurlyeq \mathcal{M}$.

Definition 26.2 A structure $\mathcal{A}$ is *atomic* if, for every finite tuple $\bar{a}$ from A, the type $\text{tp}_{\mathcal{A}}(\bar{a})$ is a principal type.

Warning The word *atomic* is related to the principal formulas which are atoms in the Boolean algebras $\text{Lind}_n(T)$; it has nothing to do with atomic formulas.

Both these definitions capture some aspect of a model being small. By the omitting types theorem, atomic models realise as few different types as possible. This gives a way to relate the two notions.

Theorem 26.3 *Let T be a complete theory in a countable language, and suppose $\mathcal{A} \models T$. Then $\mathcal{A}$ is a prime model of T if and only if $\mathcal{A}$ is countable and atomic.*

Proof By the strong version of the compactness theorem, Theorem 11.7, there is a countable model $\mathcal{B}$ of T. So for $\mathcal{A}$ to embed in $\mathcal{B}$, it must also be countable. By the omitting types theorem, given any non-principal type p, we can choose $\mathcal{B}$ to omit p. If $\mathcal{A}$ is a prime model, then by Proposition 23.4, $\mathcal{A}$ must also omit p. So $\mathcal{A}$ can only realise principal types and so is atomic. So prime models of T are countable and atomic.

For the converse direction, see Exercise 25.7. □

Theorem 26.4 *Suppose that $\mathcal{A}$ and $\mathcal{B}$ are both countable atomic models of T. Then $\mathcal{A} \cong \mathcal{B}$.*

Proof idea The proof uses the back-and-forth method and is very similar to the proof of Theorem 15.3 and (ii) ⇒ (i) from Theorem 25.4. □

Examples 26.5

(i) Every model of DLO is atomic. More generally, by the Ryll–Nardzewski theorem, every model of a countably categorical theory is atomic.
(ii) If $V \models T^{\infty}_{\mathbb{Q}\text{-VS}}$, then V is atomic if and only if $\dim V = 1$.
(iii) $\mathbb{Z}_{\text{ring}}$ is a prime model of its theory.

26.2 Saturated and Universal Models

A prime model of T embeds elementarily in every model of T. By analogy, one might define a *universal* model of T to be one in which every model of T embeds elementarily. By cardinality considerations and the Upward Löwenheim–Skolem theorem, such a model cannot exist. However, if we are more modest and merely ask for models of sufficiently small cardinality to embed, then we can hope to find such models.

Definition 26.6 A model $\mathcal{M} \models T$ is *universal* if for every model $\mathcal{A} \models T$ such that $|\mathcal{A}| \leqslant |\mathcal{M}|$, there is an elementary embedding $\mathcal{A} \preccurlyeq \mathcal{M}$.

Analogously to the situation for small models, we also consider the property of a model realising as many types as possible. However, there is a subtlety here in that we need to consider types over parameters, that is, types in expansions of T by constant symbols.

Definition 26.7 A structure $\mathcal{M}$ is $\aleph_0$-*saturated* if, for every expansion $\mathcal{M}'$ of $\mathcal{M}$ by naming finitely many elements of M by constant symbols, every n-type of $\mathrm{Th}(\mathcal{M}')$ is realised in $\mathcal{M}'$. These new constants are called *parameters*, and types of $\mathrm{Th}(\mathcal{M}')$ are called *types over parameters* of $\mathrm{Th}(\mathcal{M})$.

Theorem 26.8 *Suppose that $\mathcal{M}$ is countable and $\aleph_0$-saturated. Then $\mathcal{M}$ is a universal model of its theory.*

Proof idea The proof is very similar to Exercise 25.7. □

Theorem 26.9 *Suppose that $\mathcal{M}$ and $\mathcal{N}$ are both countable and $\aleph_0$-saturated models of the same complete theory. Then $\mathcal{M} \cong \mathcal{N}$.*

Proof idea As for the proof of Theorem 26.4, the back-and-forth method is used. □

26.3 0-Stable Theories

Not every theory has prime models or countable $\aleph_0$-saturated models. We can give a sufficient condition for both in terms of counting types.

Definition 26.10 A complete theory T is *0-stable* (zero-stable) if, for every $n \in \mathbb{N}^+$, the Stone space $S_n(T)$ is countable. Many authors say T is *small* rather than 0-stable.

By the Ryll–Nardzewski theorem, every countably categorical theory is 0-stable, since finite sets are countable.

Lemma 26.11 *Suppose T is a 0-stable L-theory and T' is a complete theory extending T in a language $L' = L \cup \{c_1, \ldots, c_r\}$ with finitely many new constant symbols. Then T' is also 0-stable.*

Proof For $n \in \mathbb{N}^+$, let f be the map $f : S_n(T') \to S_{n+r}(T)$ given by $f(p) = \{\varphi(x_1, \ldots, x_{n+r}) \mid \varphi(x_1, \ldots, x_n, c_1, \ldots, c_r) \in p\}$. It is straightforward to check that $f(p)$ really is a complete $(n + r)$-type of T and that the map f is injective. So $|S_n(T')| \leqslant |S_{n+r}(T)| \leqslant \aleph_0$. □

Theorem 26.12 *Let T be a complete theory in a countable language, L. Then T has a countable $\aleph_0$-saturated model if and only if T is 0-stable.*

Proof First suppose that T has a countable $\aleph_0$-saturated model $\mathcal{M}$. Then $\mathcal{M}$ realises every n-type in $S_n(T)$ for every $n \in \mathbb{N}^+$. Since $\mathcal{M}$ is countable, it has only countably many n-tuples. So each $S_n(T)$ must be countable.

Now suppose that T is 0-stable. Let $\mathcal{A}_1$ be any countable model of T, and expand it to $\mathcal{A}_1^+$ by naming each element $a \in A$ by a new constant symbol c_a.

For any finite subset $S \subseteq A_1$, let P_1^S be the set of all n-types (for all $n \in \mathbb{N}^+$) of the expansion of $\mathcal{A}_1$ just by the constant symbols c_a for $a \in S$. By Lemma 26.11, P_1^S is countable. There are only countably many finite subsets of A_1, so $P_1 = \bigcup \left\{ P_1^S \mid S \text{ a finite subset of } A_1 \right\}$ is also countable. By Proposition 23.7, there is a countable elementary extension $\mathcal{A}_1^+ \preccurlyeq \mathcal{A}_2^+$ which realises all the types in P_1. Then the reduct $\mathcal{A}_2$ of $\mathcal{A}_2^+$ to L is an elementary extension of $\mathcal{A}_1$. We iterate this process to get a chain of elementary extensions

$$\mathcal{A}_1 \preccurlyeq \mathcal{A}_2 \preccurlyeq \mathcal{A}_3 \preccurlyeq \cdots \preccurlyeq \mathcal{A}_r \preccurlyeq \cdots$$

indexed by natural numbers. Let $\mathcal{M}$ be the union of the chain. Then, by Exercise 13.9, $\mathcal{M}$ is an elementary extension of each $\mathcal{A}_n$. As a countable union of countable sets, it is countable. If S is any finite subset of $\mathcal{M}$, there is $r \in \mathbb{N}^+$ such that $S \subseteq A_r$. Then, by construction, $\mathcal{A}_{r+1}^+$ realises every type of T expanded by constant symbols for the elements of S. Since $\mathcal{M}$ is an elementary extension of $\mathcal{A}_{r+1}$, $\mathcal{M}^+$ also realises these types. Hence $\mathcal{M}$ is an $\aleph_0$-saturated model of T. □

There are theories with prime models which are not 0-stable. For example, $\text{Th}(\mathbb{N}_{\text{s-ring}})$ has uncountably many 1-types, but $\mathbb{N}_{\text{s-ring}}$ is a prime model. So 0-stability is not a necessary condition for the existence of prime models. However, it is sufficient.

Theorem 26.13 *Suppose T is 0-stable. Then T has a countable atomic (hence prime) model.*

Proof Let P be the set of all non-principal complete n-types of T, for all $n \in \mathbb{N}^+$. Since T is 0-stable, P is countable. Then, by Theorem 24.4, there is a countable model $\mathcal{A} \models T$ which omits all the types in P and hence is an atomic model. □

Remark 26.14 It is possible to give a necessary and sufficient condition for the existence of a prime model of T using the topology of the type spaces.

Exercises

26.1 For which K and d is a K-vector space of dimension d an atomic $L_{K\text{-vs}}$-structure? For which K and d is it prime, universal, or $\aleph_0$-saturated?

26.2 Suppose that T is a complete theory in a countable language and T has an uncountable atomic model $\mathcal{A}$. Show that T has a prime model.

26.3 Show that a complete theory T is countably categorical if and only if every model of T is atomic.

26.4 Suppose that for every expansion $\mathcal{A}'$ of $\mathcal{A}$ by finitely many constant symbols, every 1-type of $\mathrm{Th}(A')$ is realised in $\mathcal{A}'$. Show that $\mathcal{A}$ is $\aleph_0$-saturated.

26.5 Complete the proofs of all the theorems in this chapter.

26.6 A structure $\mathcal{A}$ is *strongly* $\aleph_0$*-homogeneous* if, whenever $\bar{a}$, $\bar{b}$ are n-tuples of $\mathcal{A}$ and $\mathrm{tp}_{\mathcal{A}}(\bar{a}) = \mathrm{tp}_{\mathcal{A}}(\bar{b})$, then there is an automorphism $\pi \in \mathrm{Aut}(\mathcal{A})$ such that $\pi(\bar{a}) = \bar{b}$. Using the back-and-forth method, show that countable atomic models and countable $\aleph_0$-saturated models are strongly $\aleph_0$-homogeneous.

26.7 Suppose that $\mathcal{A}$ and $\mathcal{B}$ are countable, strongly $\aleph_0$-homogeneous models of a complete theory T and they realise the same sets of n-types for all $n \in \mathbb{N}^+$. Show that $\mathcal{A} \cong \mathcal{B}$.

26.8 Suppose that $\mathcal{A}$ is countable, universal, and strongly $\aleph_0$-homogeneous. Show that $\mathcal{A}$ is $\aleph_0$-saturated. Give an example of a countable universal structure which is not $\aleph_0$-saturated.

26.9 Let P be the set of all prime numbers in $\mathbb{N}$. Given any subset Q of P, show there is a 1-type of $\mathbb{N}_{\text{s-ring}}$ of an element which is divisible by all the primes in Q but no other primes in P. Deduce that $S_1(\mathbb{N}_{\text{s-ring}})$ is uncountable.

26.10 Suppose that $\mathcal{M}$ is countable and $\aleph_0$-saturated and that $S \subseteq M^n$. Show that S is preserved under all automorphisms of $\mathcal{M}$ if and only if S is a union of type-definable sets.

27
Saturated Models

The only method we have used in this book to prove that a given axiomatisation of a theory is complete is the Łos–Vaught test: a theory with no finite models which is categorical in some infinite cardinality is complete. In Chapter 18 we briefly discussed how the back-and-forth method can be adapted to give a second method. In this chapter we give a third method using saturated models and apply it to the theory of discrete linear orders without endpoints. This works in the same spirit as the Łos–Vaught test, but we need to use more set theory than before: at least transfinite induction, and for some theories, more.

27.1 Saturation

Theorem 26.9 is suggestive. It states that countable $\aleph_0$-saturated models of a complete theory are unique up to isomorphism. Suppose a theory T has no finite models and we can prove that any two countable $\aleph_0$-saturated models of T are isomorphic. Then we want to conclude that T is complete. The problem is that T might not have any countable $\aleph_0$-saturated model or that it might have a completion with no countable $\aleph_0$-saturated model. So we would still need to prove that every completion of T is 0-stable to deduce that T is complete. We can address this problem by generalising the notion of $\aleph_0$-saturation.

Definition 27.1 Let κ be an infinite cardinal. A structure $\mathcal{M}$ is *κ-saturated* if, for every subset $A \subseteq M$ such that $|A| < \kappa$, writing $\mathcal{M}_A$ for the expansion of $\mathcal{M}$ by constant symbols for the elements of A, every n-type of $\mathrm{Th}(\mathcal{M}_A)$ is realised in $\mathcal{M}_A$.

$\mathcal{M}$ is *saturated* if it is $|M|$-saturated.

As in Exercise 26.4, it is sufficient to consider 1-types rather than n-types.

If we use the same back-and-forth proof as for Theorem 26.9, but with transfinite induction in place of induction on natural numbers, we get the following.

Theorem 27.2 *Suppose $\mathcal{A}$, $\mathcal{B}$ are saturated models of the same complete theory T, of the same cardinality. Then $\mathcal{A} \cong \mathcal{B}$.* □

27.2 Stability

We still have the problem of the existence of these saturated models. As in the countable case, the only obstacle is that there may be too many types to fit into a model, without making the model too large. There is an analogue of 0-stability.

Definition 27.3 Let κ be an infinite cardinal. A complete L-theory T is κ-*stable* if, over any set of at most κ parameters in any model of T, there are at most κ complete types.

T is *stable* if it is κ-stable for some infinite κ. Otherwise, it is *unstable*.

If T is κ-stable, then a transfinite version of the construction for the proof of Theorem 26.12 shows that there is a saturated model of T of cardinality κ.

Surprisingly, stability is a much more robust notion than 0-stability, and it has several other equivalent definitions which, on the face of it, are unrelated to counting types. For example:

Fact 27.4 *Suppose $\mathcal{M}$ is a structure in which some infinite linear order is definable. Then* Th($\mathcal{M}$) *is unstable.*

A related but more complicated condition called the *Order Property* is in fact equivalent to instability. Such conditions sometimes make it possible to prove that the theory of a structure is stable without first proving the completeness of any axiomatisation.

Stability theory and its generalisations are a major area of research in model theory. The beginnings of the subject are covered in Marker [Mar02], Tent and Ziegler [TZ12], and Baldwin [Bal88]. Poizat [Poi00] and Pillay [Pil83] offer a different approach. More advanced books taking the subject in different directions are Pillay's [Pil96], which highlights the connections between stability theory and geometric ideas, and Shelah's great work [She90], which is still the best guide to the subject for those with the perseverance to work through it.

27.3 Strongly Inaccessible Cardinals

If a theory is unstable, which by Fact 27.4 the theory of discrete linear orders is, we can find a saturated model by working harder with the set theory.

Given a subset A of a model of an L-theory T with $|A| = \kappa$, we have $|\operatorname{Form}(L_A)| = \kappa + \aleph_0$. A type over A is a subset of $\operatorname{Form}(L_A)$ so there are at most $|\mathcal{P}\operatorname{Form}(L_A)|$ types.

Definition 27.5 A cardinal κ is *strongly inaccessible* if, for a set X of size κ, whenever $Y \subseteq X$ with $|Y| < \kappa$, then also $|\mathcal{P}Y| < \kappa$, and furthermore, if $(Y_i)_{i \in I}$ is a family of subsets of X such that $|I| < \kappa$ and $|Y_i| < \kappa$ for each $i \in I$, then $\bigcup_{i \in I} Y_i \neq X$.

Again, a transfinite version of the proof of Theorem 26.12 gives us saturated models.

Theorem 27.6 *Let T be an L-theory with infinite models, and let κ be a strongly inaccessible cardinal such that $\kappa > |L|$. Then T has a saturated model of cardinality κ.* □

So we have the following analogue of the Łos–Vaught test.

Theorem 27.7 *Suppose T is an L-theory with no finite models and κ is a strongly inaccessible cardinal such that $\kappa > |L|$. Then T is complete if and only if any two saturated models of T of cardinality κ are isomorphic.*

Proof Combine Theorems 27.2 and 27.6. □

27.4 Discrete Linear Orders

We give an application of Theorem 27.7.

Theorem 27.8 *The theory* DiscLO *of nonempty discrete linear orders without endpoints is complete.*

Proof From Exercise 15.8, any discrete linear order without endpoints is isomorphic to a lexicographic product $\mathcal{A} \times \mathbb{Z}_<$, where $\mathcal{A}$ is a non-empty linear order. Conversely, for any non-empty linear order $\mathcal{A}$ it is easy to see that $\mathcal{A} \times \mathbb{Z}_< \models \text{DiscLO}$.

Suppose that $\mathcal{M} = \mathcal{A} \times \mathbb{Z}_<$ is saturated (or even just $\aleph_0$-saturated). By Exercise 15.2, the function s taking an element of $\mathcal{M}$ to its immediate successor is definable, and likewise the predecessor function s^{-1} is definable. Given $\alpha \in \mathcal{M}$, consider the set of formulas with parameter α given by

$\{s^n(x) < \alpha \mid n \in \mathbb{N}^+\}$. This is finitely satisfiable, so a 1-type, so by saturation, it is realised in $\mathcal{M}$. Thus $\mathcal{A}$ has no least endpoint. Similarly, $\mathcal{A}$ has no greatest endpoint. Given $a < b$ in $\mathcal{A}$, we have $\alpha = (a, 0)$ and $\beta = (b, 0)$ in $\mathcal{M}$. Then $\{\alpha < s^{-n}(x) \wedge s^n(x) < \beta \mid n \in \mathbb{N}^+\}$ is a 1-type over the parameter set $\{\alpha, \beta\}$ and so is realised in $\mathcal{M}$. Thus $\mathcal{A}$ is a dense linear order. Using quantifier elimination for DLO, we can understand that 1-types of DLO over any set of parameters just correspond to cuts in the order. It follows that since $\mathcal{M}$ is saturated, so is $\mathcal{A}$.

Now suppose that $\mathcal{M} = \mathcal{A} \times \mathbb{Z}_<$ and $\mathcal{N} = \mathcal{B} \times \mathbb{Z}_<$ are two saturated models of DiscLO of the same strongly inaccessible cardinality κ. Then $\mathcal{A}$ and $\mathcal{B}$ are saturated models of DLO of cardinality κ. Since DLO is a complete theory, $\mathcal{A} \cong \mathcal{B}$. Thus $\mathcal{M} \cong \mathcal{N}$. So by Theorem 27.7, DiscLO is complete. □

27.5 Discussion

Note that the proof above of Theorem 27.8 is almost entirely about understanding the models of the theory, in fact, countable parts of models, and relating them to something we already understand (countable parts of models of DLO). The set theory behind the saturated models is not visible. Any proof that DiscLO is complete needs to use at least some comparable level of understanding of the models. In practice, this is exactly how saturated models are used to simplify proofs, which could in principle be done by working entirely inside small, often countable, models.

We have focused on proving completeness of theories here, but another advantage of saturated models is that they have a lot of automorphisms, so the methods we have used for quantifier elimination and for understanding definable sets or types more generally can be extended easily.

One minor disadvantage to the method of saturated models, at least for unstable theories, is the requirement to work with strongly inaccessible cardinals. Under the usual set theoretic axioms (ZFC), you cannot prove that strongly inaccessible cardinals do or do not exist. There are a number of different ways to get around this. You can just work in this slightly stronger set theory with strongly inaccessible cardinals or assume the generalised continuum hypothesis, which gives another reason for saturated models to exist. Then you can use set-theoretic methods to show that the theorems you prove with these extra assumptions could also be proven in ZFC. Chang and Keisler [CK90] use *special models*, which are similar to saturated models but do always exist, and (for countable recursive theories) they also use *recursively saturated models*. Tent and Ziegler [TZ12] and Hodges [Hod93] take different approaches again.

Exercises

27.1 Show that $\mathbb{R}_<$ is not a saturated model of DLO.

27.2 Show that DiscLO is 0-stable. What is its countable saturated model?

27.3 Let $\mathcal{M} \models$ DLO and let $\mathcal{A}$ be a substructure with domain A. Using quantifier elimination for DLO, show that 1-types of $\mathcal{M}$ over A correspond to cuts in the order of $\mathcal{A}$, that is, subsets $C \subseteq A$ such that if $c \in C$ and $a \in A$ with $a < c$, then $a \in C$.

27.4 Sketch a proof of Theorem 27.8, including any theorems and lemmas which are used.

27.5 Show that, for each $n \in \mathbb{Z}$, the function f_n given by $x \mapsto x + n$ is definable in $\mathbb{Z}_<$, and hence in any model of DiscLO. Then show that $\langle \mathbb{Z}; <, (f_n)_{n\in\mathbb{Z}} \rangle$ has quantifier elimination.

27.6 Characterise the models of the theory of infinite discrete linear orders with endpoints, and show the theory is complete.

27.7 Suppose we alter Definition 27.3 to say that T is κ-stable if over any set of at most κ parameters there are at most $\kappa + \aleph_0$ types, and we remove the condition that κ be infinite. Show that the new definition is equivalent to the old one when κ is infinite and that for any $n \in \mathbb{N}$ the newly defined notion of n-stable is equivalent to Definition 26.10 of 0-stable.

27.8 This exercise, analogous to Exercise 26.6 in the countable case, indicates how saturated structures have a lot of automorphisms. A structure $\mathcal{A}$ is *strongly κ-homogeneous* if, whenever $\bar{a}$ and $\bar{b}$ are potentially infinite tuples of length strictly less than κ such that $\text{tp}(\bar{a}) = \text{tp}(\bar{b})$, there is an automorphism π of $\mathcal{A}$ such that $\pi(\bar{a}) = \bar{b}$. Make sense of this definition, and use transfinite induction to prove that a saturated structure of cardinalty κ is strongly κ-homogeneous.

27.9 Referring to another source, such as [Poi00], [Mar02], or [TZ12], use the back-and-forth method to give another proof that DiscLO is complete.

27.10 Assuming the generalised continuum hypothesis, show that every theory has saturated models.

Part VI

Algebraically Closed Fields

One of the largest areas of applications of model theory is to the model theory of fields. These may be fields considered just in the ring language or fields with operators such as derivations, automorphisms, or valuations; fields equipped with certain analytic functions; or fields equipped with relation symbols naming certain subgroups or subfields. In many cases the fields under consideration are algebraically closed, but even when they are not, the theory of algebraically closed fields usually plays a role. In this final part of the book we will study the theory of algebraically closed fields. Two chapters give the basic algebra of fields and of algebraic closures, then we prove categoricity, completeness, and quantifier elimination by the methods developed in Parts III and IV, in particular the Łos–Vaught test and substructure completeness. We show how definable sets are algebraic varieties and give a model-theoretic proof of Hilbert's Nullstellensatz. The connection between the Stone spaces of types and the Zariski spectra of prime ideals is given as an exercise.

In this part of the book, every structure will be a ring considered as an L_{ring}-structure, so rather than writing a ring as R_{ring} or $\mathcal{R}$, we will use the same symbol R for both a ring and its underlying set.

28

Fields and Their Extensions

In this chapter we review the basic concepts of field extensions we need, up to the Hilbert basis theorem. Details and proofs can be found in basic textbooks on rings and fields, and Lang [Lan02] is a standard reference. To understand fields with extra structure, one usually has to develop or understand the analogous concepts.

28.1 The Basic Algebra of Rings

Recall the language of rings is $L_{\text{ring}} = \langle +, \cdot, -, 0, 1 \rangle$. The axioms for rings and for fields were given in Chapter 6.

Definition 28.1 A *domain* (sometimes called an *integral domain*) is a ring with no zero divisors, that is, it is a model of the sentence

$$\forall xy[x \cdot y = 0 \rightarrow (x = 0 \lor y = 0)].$$

Every field is a domain, and since the ring axioms and the property of being a domain are given by universal sentences, every L_{ring}-substructure of a field is a domain. The converse is also true: every domain is a substructure of a field, in particular its field of fractions. The following lemma captures the property we need.

Lemma 28.2 *If R is a domain, then it has a* field of fractions, *which is a field $F_R \supseteq R$ such that any embedding $\pi : R \hookrightarrow F$ of R into a field F extends uniquely to an embedding $F_R \hookrightarrow F$.*

Examples 28.3 We write $R[x_1, \dots, x_n]$ for the ring of polynomials in the variables $x_1, \dots, x_n$, with coefficients from a ring R. If R is a domain, so is $R[x_1, \dots, x_n]$. If R is a field F, the field of fractions of $F[x_1, \dots, x_n]$ is written as

$F(x_1, \ldots, x_n)$. It is called the field of *rational functions* over F in the variables $x_1, \ldots, x_n$.

Definition 28.4 A *homomorphism* of rings $\pi : R \to S$ is a function that preserves $+, \cdot, -, 0$, and 1. The *kernel* of π is $\ker(\pi) = \{a \in R \mid \pi(a) = 0\}$.

An *ideal* of a ring R is an additive subgroup $I \subseteq R$ such that, for all $r \in R$ and all $a \in I$, we have $r \cdot a \in I$.

The ideal *generated* by a subset $A \subseteq R$ is the intersection of the ideals of R which contain A.

Lemma 28.5 *Suppose $I \subseteq R$ is the ideal generated by $f_1, \ldots, f_r$, and $f \in R$. Then $f \in I$ if and only if there are $g_1, \ldots, g_r \in R$ such that $f = \sum_{i=1}^{r} g_i f_i$.*

Lemma 28.6 *The kernel of a ring homomorphism $\pi : R \to S$ is an ideal of R.*

If I is an ideal of R, there is a quotient ring *R/I whose elements are the cosets $a + I = \{a + b \mid b \in I\}$ for $a \in R$. The map $\pi : R \to R/I$ defined by $\pi(a) = a + I$ is a surjective ring homomorphism with kernel I.*

Definition 28.7 An ideal I of R is *proper* if $I \neq R$. It is *prime* if it is proper and for all $a, b \in R$, if $ab \in I$, then $a \in I$ or $b \in I$. An ideal I is *maximal* if it is proper and there is no proper ideal J of R with $I \subsetneq J$.

Lemma 28.8 *Let R be a ring and I an ideal of R. Then the quotient R/I is a domain if and only if I is prime, and it is a field if and only if I is maximal.*

Example 28.9 The ring $\mathbb{Z}/n\mathbb{Z}$ is a domain if and only if n is a prime p, and then it is a field written $\mathbb{F}_p$.

Definition 28.10 The *characteristic* of a field F is the smallest $p \in \mathbb{N}^+$ such that $\underbrace{1 + \cdots + 1}_{p} = 0$ if such a p exists, and 0 otherwise.

The *prime subfield* of a field F is the intersection of all the subfields of F.

If F has characteristic $p > 0$, then the prime subfield of F is $\mathbb{F}_p$ and if F has characteristic 0, then its prime subfield is the field of rational numbers $\mathbb{Q}$.

28.2 Simple Field Extensions

Algebraically closed fields arise as extensions of any field. To understand this, we need to understand the finitely generated field extensions. We start with the extensions generated by just one element.

Definition 28.11 A *simple extension* of a field F is a field extension K generated by a single element, say, b. Then we write $K = F(b)$. We write $F[b]$ for the subring of K generated by F and b. So $F(b)$ is the field of fractions of $F[b]$.

Lemma 28.12 *Let $F \subseteq F(b)$ be a simple field extension. There is a unique homomorphism* $\mathrm{ev}_b : F[x] \to F(b)$ *such that* $\mathrm{ev}_b(a) = a$ *for all* $a \in F$ *and* $\mathrm{ev}_b(x) = b$.

The kernel of ev_b is $I_b := \{f(x) \in F[x] \mid f(b) = 0\}$, and it determines the field extension $F(b)$ of F and the choice of generator up to isomorphism. So to classify the simple extensions of F, it is enough to classify the prime ideals of $F[x]$.

Proposition 28.13 *Every ideal of $F[x]$ is generated by a single polynomial, which is unique up to multiplication by a non-zero element of F. (That is, $F[x]$ is a* principal ideal domain.*) The ideal is prime if and only if the generating polynomial is* irreducible.

A generating polynomial $g(x)$ for the ideal I_b is called a *minimal polynomial* for b over F. We write the ideal I_b as $\langle g \rangle$. So simple extensions of a field F are determined up to isomorphism fixing a choice of generator, by the minimal polynomial of the generator, which is an irreducible polynomial in $F[x]$.

Examples 28.14 Suppose $F = \mathbb{Q}$ and $b = \sqrt{2} \in \mathbb{Q}(\sqrt{2}) \subseteq \mathbb{C}$. Then the minimal polynomial of $\sqrt{2}$ is $x^2 - 2$. The same field $\mathbb{Q}(\sqrt{2})$ is also generated by $c = \sqrt{2} + 1$. The minimal polynomial of $\sqrt{2} + 1$ is $x^2 - 2x - 1$.

The number π is *transcendental*, that is, it does not satisfy any polynomial equations with coefficients from $\mathbb{Q}$. So its minimal polynomial over $\mathbb{Q}$ is the zero polynomial, and the ideal I_π is the zero ideal $\{0\}$. Then the quotient map $\mathrm{ev}_\pi : \mathbb{Q}[x] \to \mathbb{Q}(\pi)$ is injective, and the field $\mathbb{Q}(\pi)$ is isomorphic to the field $\mathbb{Q}(x)$ of rational functions in one variable over $\mathbb{Q}$.

28.3 Finitely Generated Field Extensions

If $F \subseteq K$ is a field extension generated by finitely many elements, say, $\bar{b} = (b_1, \ldots, b_n)$, then we write $K = F(\bar{b})$ as in the simple case. Again, K is the field of fractions of the quotient of $F[\bar{x}]$ by the ideal $\left\{f(\bar{x}) \in F[\bar{x}] \;\middle|\; f(\bar{b}) = 0\right\}$, which is a prime ideal. So again, to classify the finitely generated field extensions of a field F, it is enough to classify the prime ideals of $F[x_1, \ldots, x_n]$, for each

$n \in \mathbb{N}$. For $n > 1$, these are not always generated by a single polynomial, but we have the next best thing.

Theorem 28.15 (Hilbert's basis theorem) *For any ideal I of $F[x_1, \ldots, x_n]$, there is a finite set of polynomials which generates I.*

Definition 28.16 If $F \subseteq K$ is a field extension and $b \in K$, then b is said to be *algebraic* over F if b satisfies some non-zero polynomial over F. Otherwise, it is *transcendental* over F. If every element of K is algebraic over F, then K is said to be an algebraic extension of F.

Finitely generated algebraic extensions have a special form.

Proposition 28.17 *If $K = F(\bar{b})$ is a finitely generated algebraic extension of F, then there is a single element a which generates K as an extension of F. Furthermore, the ring $F[a]$ is actually all of K.*

Finally, we give two useful lemmas.

Lemma 28.18 *If $F \subseteq K \subseteq L$ is a tower of field extensions, L is algebraic over K, and K is algebraic over F, then L is algebraic over F.*

A *zero* of a polynomial $f(x)$ is a solution to the equation $f(x) = 0$.

Lemma 28.19 *Let F be a field and $f(x) \in F[x]$ a polynomial in one variable, of degree d. Then there are at most d zeros of f in F.*

This chapter is intended as a reminder or reference, but it covers the material rather quickly to learn it from. Thus there are no exercises, other than to read up on any parts of the chapter you are unfamiliar with from a suitable reference and to do the exercises from there.

29
Algebraic Closures of Fields

In this chapter we study algebraically closed fields from an algebraic point of view. In particular, we prove that every field has an algebraic closure, which is unique up to isomorphism.

There are actually two different notions called algebraic closure relating to fields. They are closely related, and often the difference is not made explicit, but they play different roles in the model-theoretic approach. Both roles are important.

29.1 Relative Algebraic Closure

Definition 29.1 Suppose that $F \subseteq K$ is an extension of fields. The *relative algebraic closure* of F in K, written $\mathrm{acl}^K(F)$, consists of all the elements of K which are algebraic over F.

For example, $\mathrm{acl}^{\mathbb{R}}(\mathbb{Q})$ is the field of real algebraic numbers. It is not an algebraically closed field because it is a subfield of $\mathbb{R}$ and so, for example, the polynomial $x^2 + 1$ has no zero in $\mathrm{acl}^{\mathbb{R}}(\mathbb{Q})$.

Lemma 29.2 *For any extension $F \subseteq K$ of fields, $\mathrm{acl}^K(F)$ is a subfield of K of cardinality $|F| \leqslant |\mathrm{acl}^K(F)| \leqslant \max\{|F|, \aleph_0\}$. In particular, if F is an infinite field, then $|\mathrm{acl}^K(F)| = |F|$.*

Proof We leave the proof that $\mathrm{acl}^K(F)$ is a subfield as an exercise. For the cardinality result, note that each element of $\mathrm{acl}^K(F)$ is a zero of a polynomial $f(x) \in F[x]$, and f has at most $\deg(f)$ zeros, a finite number. So $|\mathrm{acl}^K(F)| \leqslant |F[x]| \cdot \aleph_0$. But $|F[x]| = \max\{|F|, \aleph_0\}$, so $|\mathrm{acl}^K(F)| \leqslant \max\{|F|, \aleph_0\}$. Also $F \subseteq \mathrm{acl}^K(F)$, so $|F| \leqslant |\mathrm{acl}^K(F)|$. □

29.2 (Absolute) Algebraic Closure

Recall that a field F is *algebraically closed* if every non-constant polynomial $f(x) \in F[x]$ has a zero in F. A first-order axiom scheme capturing this property was given in Chapter 7. The fundamental theorem of algebra states that the complex field $\mathbb{C}_{\text{ring}}$ is algebraically closed. Now we show that there are many other algebraically closed fields.

Proposition 29.3 *For any field F, there is an extension $F \subseteq K$ such that K is an algebraically closed field.*

We give a proof using the compactness theorem.

Proof Expand L_{ring} to L_F by adding constant symbols naming each element of F. Recall from Chapter 13 that we have $\text{Diag}(F)$, the set of all atomic L_F-sentences and negations of atomic L_F-sentences which are true in F. Let $\Sigma = \text{Diag}(F) \cup \{\exists x[f(x) = 0] \mid f(x) \in F[x] \text{ and } f(x) \text{ is non-constant}\}$. Then a model of Σ is an extension field of F in which every non-constant polynomial over F has a zero. Let Σ_0 be a finite subset of Σ, and let $\{f_1(x), \ldots, f_r(x)\}$ be the finite set of polynomials $f(x)$ such that the axiom $\exists x[f(x) = 0]$ appears in Σ_0. Let $g_1(x)$ be an irreducible component of $f_1(x)$, and let $F_1 = F[x]/\langle g_1 \rangle$. Now let $g_2(x)$ be an irreducible component of $f_2(x)$ considered as a polynomial over F_1, and let $F_2 = F_1[x]/\langle g_2 \rangle$. Iterating, we get a field extension F_r of F in which all the polynomials $f_1(x), \ldots, f_r(x)$ have zeros. Thus Σ is finitely satisfiable. By compactness, Σ has a model, say, K_1.

Now iterate the process to get a chain of field extensions

$$F = K_0 \subseteq K_1 \subseteq K_2 \subseteq \cdots \subseteq K_n \subseteq \cdots$$

for $n \in \mathbb{N}$ such that every non-constant $f \in K_n[x]$ has a zero in K_{n+1}. Let $K = \bigcup_{n \in \mathbb{N}} K_n$. Then K is an algebraically closed field extension of F. □

Now we come to the second notion of algebraic closure of F. Unlike the relative algebraic closure, it is defined without reference to a previously given extension field K.

Definition 29.4 An *algebraic closure* of a field F is an algebraic field extension of F which is algebraically closed. Very rarely, this notion is called the *absolute algebraic closure of F* to distinguish it from the relative algebraic closure. Usually we let the context make the distinction.

Putting together the previous proposition with the relative algebraic closure, we can show the existence of an algebraic closure of a field F. With some more work, we can also show uniqueness of the algebraic closure.

Theorem 29.5 *Every field F has an algebraic closure. If K and L are both algebraic closures of F, then there is an isomorphism $K \cong L$ which restricts to the identity on F.*

Proof Let F be any field. By Proposition 29.3, there is an algebraically closed field extension K of F. Take $K_0 = \mathrm{acl}^K(F)$. Suppose that $f(x)$ is a non-constant polynomial in $K_0[x]$. Then since K is algebraically closed, there is $b \in K$ such that $f(b) = 0$. Then b is algebraic over K_0, which is algebraic over F, so, by Lemma 28.18, b is algebraic over F, and hence $b \in K_0$. So K_0 is an algebraically closed field which is an algebraic extension of F. So we have shown that F has at least one algebraic closure.

Now suppose that K and L are both algebraic closures of F. We must find an isomorphism $\pi : K \to L$ fixing F pointwise. We give a proof using transfinite induction.

List K as $(a_\alpha)_{\alpha<\lambda}$ for some ordinal λ. (If K is countable, then just take the α to be natural numbers.) We define subfields K_α of K and embeddings $\pi_\alpha : K_\alpha \to L$ for $\alpha \leqslant \lambda$ as follows:

- $K_0 = F$ and $\pi_0 : F \to L$ is just the inclusion of F as a subfield of L.
- If α is a limit ordinal, take $K_\alpha = \bigcup_{\beta<\alpha} K_\beta$ and $\pi_\alpha = \bigcup_{\beta<\alpha} \pi_\beta$.
- If $\alpha = \beta + 1$ is a successor ordinal, let $K_\alpha = K_\beta(a_\beta)$. Let $f(x)$ be the minimal polynomial of a_β over K_β. Then since a_β is algebraic over F (and hence over K_α), we have $K_\alpha \cong K_\beta[x]/\langle f\rangle$. Let $L_\beta = \pi_\beta(K_\beta)$, a subfield of L, and let $g(x) \in L_\beta[x]$ be the polynomial obtained from $f(x)$ by applying π_β to all the coefficients. Since L is algebraically closed, $g(x)$ has a zero in L. Let b be such a zero. Since $f(x)$ is irreducible in $K_\beta[x]$, $g(x)$ is irreducible in $L_\beta[x]$, so it is the minimal polynomial of b over L_β. So we have

$$K_\beta(a_\beta) \cong K_\beta[x]/\langle f\rangle \cong L_\beta[x]/\langle g\rangle \cong L_\beta(b) \subseteq L$$

  via a map taking a_β to b, which extends π_β. Let π_α be this embedding of K_α into L.

Let $\pi = \pi_\lambda$. Then π is an embedding of K into L, which extends π_0, the identity map on F. It remains to show that π is surjective.

Let $b \in L$ and let $f(x)$ be its minimal polynomial over F. By Lemma 28.19, $f(x)$ has only finitely many zeros in K, say, $a_1, \ldots, a_n$. Since π is an L_{ring} embedding fixing F, we have $f(\pi(a_i)) = 0$ for each $i = 1, \ldots, n$, and the $f(a_i)$ are all distinct because f is injective. So L has at least n zeros of f. But we can swap the roles of K and L, and the above argument shows there is also an embedding of L into K. So L has at most n zeros of f. Hence L has exactly

n zeros of f, so for some i, $f(a_i) = b$. So π is surjective, and hence it is an isomorphism. □

We usually write F^{alg} for the the algebraic closure of F.

Exercises

29.1 Show that if F is an algebraically closed field and $f \in F[x]$ is a non-zero irreducible polynomial, then f has the form $f(x) = x - a$ for some $a \in F$. Deduce that an algebraically closed field has no proper algebraic extensions.

29.2 Show that for any field F, there is no algebraically closed proper subfield of F^{alg} which contains F.

29.3 Let $F \subseteq K$ be an extension of fields. Explain how K can be regarded as an F-vector space. Let $\alpha \in K$. Show that $\alpha \in \mathrm{acl}^K(F)$ if and only if the field $F(\alpha)$ is finite-dimensional as an F-vector space.

29.4 For p prime, show that $\mathbb{F}_p^{\mathrm{alg}}$ is the union of its finite subfields.

29.5 Let $F \subseteq K$ be an extension of fields. Prove that $\mathrm{acl}^K(F)$ is a subfield of K.

29.6 The use of the compactness theorem in the proof of Proposition 29.3 is actually hiding a use of the axiom of choice or transfinite induction. Give a proof not using compactness when the field F is countable. Where is the axiom of choice used in the proof of the compactness theorem?

29.7 Give another proof of the uniqueness part of Theorem 29.5 using the back-and-forth method, avoiding a separate proof that the function π is surjective.

30
Categoricity and Completeness

In this chapter we prove that the theory of algebraically closed fields of a given characteristic is categorical in uncountable cardinalities, even after adding parameters for a subfield. We deduce that the theory is complete and has quantifier elimination. The new algebraic concept we use is that of a transcendence base, which is an analogue of a basis for a vector space.

30.1 Transcendence Bases

The notions of a basis of a vector space and its dimension have analogues for fields, which are called transcendence base and transcendence degree.

Definition 30.1 Let $F \subseteq K$ be a field extension. A subset B of K is said to be *algebraically independent over* F if, for every non-zero polynomial $f(x_1, \ldots, x_n) \in F[x_1, \ldots, x_n]$, for every $n \in \mathbb{N}$, if $b_1, \ldots, b_n \in B$ are distinct, then $f(b_1, \ldots, b_n) \neq 0$.

A *transcendence base* for the extension $F \subseteq K$ is a subset B of K which is algebraically independent over F and such that K is algebraic over the subfield generated by $F \cup B$.

Lemma 30.2 *Let $F \subseteq K$ be a field extension, and let $B \subseteq K$ be any subset. Let $F[(x_b)_{b \in B}]$ be the ring of polynomials over F with one variable x_b for each element $b \in B$. Consider the ring homomorphism $\mathrm{ev}_B : F[(x_b)_{b \in B}] \to K$, which is the identify on F and satisfies $\mathrm{ev}_B(x_b) = b$. Then B is algebraically independent over F if and only if the map ev_B is injective.*

Proof By definition, B is algebraically independent over F if and only if the kernel of ev_B is the zero ideal if and only if ev_B is injective. □

Proposition 30.3 *Let K be an infinite field, let $F \subseteq K$ be a subfield, and let B be a transcendence base for K over F. Then $|K| = \max\{|F|, |B|, \aleph_0\}$.*

Proof We write $F(B)$ for the subfield of K generated by $F \cup B$.

Then we have $F(B) \subseteq K \subseteq K^{\text{alg}}$, and $\text{acl}^{K^{\text{alg}}}(F(B)) = K^{\text{alg}}$. So by Lemma 29.2, we have $|F(B)| \leqslant |K^{\text{alg}}| = |K|$, with equality if $F(B)$ is infinite. If F is a finite field and $B = \emptyset$, then $F(B) = F$ is finite, so $|K^{\text{alg}}| = \aleph_0$, and we are done. Otherwise, $F(B)$ is infinite and $|F(B)| = \max\{|F|, |B|, \aleph_0\}$, so we are done. □

Proposition 30.4 *Any two transcendence bases of a field extension $F \subseteq K$ have the same cardinality.*

Proof We leave the proof as an exercise, except in the case when the cardinality of some transcendence base B is greater than $|F| + \aleph_0$. In that case, for any transcendence base B', we have $|B'| = |K| = |B|$ by Proposition 30.3. □

Definition 30.5 We define the *transcendence degree* of a field extension to be the cardinality of any transcendence base. The *transcendence degree* of a field F is the transcendence degree of F considered as an extension of its prime subfield.

30.2 Categoricity

Recall from Chapter 7 that we write ACF_p for the theory of algebraically closed fields of characteristic p, axiomatised by the axioms for ACF and either the sentence χ_p given by $\underbrace{1 + \cdots + 1}_{p} = 0$ or, if $p = 0$, the set of sentences $\left\{\neg\chi_q \mid q \in \mathbb{N}^+\right\}$.

Theorem 30.6

(i) *For each p, prime or 0, the theory* ACF_p *is categorical in all uncountable cardinals.*
(ii) *For any field F, the theory* $\text{Diag}(F) \cup \text{ACF}$ *is categorical in all uncountable cardinals λ such that $\lambda > |F|$.*

Proof First we deduce (i) from (ii). For p prime, $K \models \text{Diag}(\mathbb{F}_p) \cup \text{ACF}$ if and only if K is an algebraically closed field of characteristic p, with constant symbols naming $0, 1, 2, \ldots, p-1$, if and only if K (without the extra constant symbols) is a model of ACF_p. For $p = 0$, the same applies with $\mathbb{Q}$ in place of $\mathbb{F}_p$. So (i) follows from (ii).

Now we prove (ii). Suppose that K and L are both models of Diag(F) $\cup$ ACF of the same uncountable cardinality λ, with $\lambda > |F|$. Then K and L are both algebraically closed field extensions of F, and by Proposition 30.3, the transcendence degrees of K and L as extensions of F are both λ. So let B be a transcendence base for $F \subseteq K$, and let B' be a transcendence base for $F \subseteq L$. Then there is an isomorphism π between the subfields $F(B)$ of K and $F(B')$ of L, restricting to the identity on F, since by Lemma 30.2, they are both isomorphic to the field of rational functions in λ variables. Then, by Theorem 29.5, this isomorphism π extends to an isomorphism $\pi' : K \to L$. So the theory Diag(F) $\cup$ ACF is categorical in λ. □

Corollary 30.7 *For each p, prime or 0, the theory* ACF_p *is complete and has quantifier elimination.*

Proof By the Łos–Vaught test, Lemma 14.11, it follows from part (i) of Theorem 30.6 that the theories ACF_p are complete.

Applying the Łos–Vaught test to part (ii) of the same theorem, we deduce that ACF is substructure complete. By Proposition 18.2, it has quantifier elimination, and thus so does each completion ACF_p. □

Putting together results from this chapter, we have the following classification of algebraically closed fields.

Theorem 30.8 *For each cardinal λ and each p, either a prime number or 0, up to isomorphism, there is exactly one algebraically closed field of characteristic p and transcendence degree λ.* □

Exercises

30.1 Prove that the characteristic of any field is either 0 or a prime number. Deduce that ACF_0 is axiomatised by the axioms for ACF together with $\{\neg\chi_p \mid p \text{ is prime}\}$.

30.2 Use the compactness theorem to deduce that if σ is any L_{ring}-sentence, then $\mathrm{ACF}_0 \vdash \sigma$ if and only if there is $N \in \mathbb{N}$ such that for all primes $p > N$, $\mathrm{ACF}_p \vdash \sigma$.

30.3 Given a set X, a function $\mathrm{cl} : \mathcal{P}X \to \mathcal{P}X$ is a *closure operator* if, for all $A, B \subseteq X$ and all $a, b \in X$: if $A \subseteq B$, then $\mathrm{cl}(A) \subseteq \mathrm{cl}(B)$, $A \subseteq \mathrm{cl}(A)$, and $\mathrm{cl}(\mathrm{cl}(A)) = \mathrm{cl}(A)$.

A closure operator cl has *finite character* if, whenever $a \in \mathrm{cl}(A)$, then there is a finite subset $A_0 \subseteq A$ such that $a \in \mathrm{cl}(A_0)$.

A closure operator cl has the *exchange property* if, whenever $b \in \text{cl}(A \cup \{a\}) \setminus \text{cl}(A)$, then $a \in \text{cl}(A \cup \{b\})$.

A *pregeometry* is a closure operator with finite character which satisfies the exchange property.

Show that acl^K is a pregeometry on any field K and that that linear span is a pregeometry on any vector space.

30.4 Using the exchange property for acl^K, prove Proposition 30.4.

30.5 Give definitions of *independent set*, *spanning set*, and *basis* for a pregeometry, by analogy to those for vector spaces. Prove that every pregeometry has a basis and that any two bases have the same cardinality.

30.6 Give an outline of the proof that the theory ACF_0 is complete and has quantifier elimination from first principles, that is, including all the main results in this book which are used in the proof.

30.7 We outline a proof due to Ax [Ax69] of a theorem of Bailynicki-Birula and Rosenlicht [BBR62]. In Exercise 31.10, we will give Ax's more general theorem.

If $f_1, \dots, f_n \in K[x_1, \dots, x_n]$, then $f = (f_1, \dots, f_n)$ defines a polynomial map $K^n \to K^n$. Let $\Psi(K, n, d)$ denote the statement 'for all such f, if each f_i has degree at most d and f is injective then f is surjective.'

(a) Show that $\Psi(K, n, d)$ holds when K is a finite field.
(b) Using Exercise 29.4, show that $\Psi(\mathbb{F}_p^{\text{alg}}, n, d)$ is true for p prime.
(c) By quantifying over the coefficients of the polynomials, for each fixed n and d, show that there is an L_{ring}-sentence $\psi_{n,d}$ such that for any field K, $K \models \psi_{n,d}$ if and only if $\Psi(K, n, d)$ is true.
(d) Using the completeness of ACF_p, deduce that $\Psi(K, n, d)$ is true for all algebraically closed fields of positive characteristic.
(e) Deduce that $\Psi(K, n, d)$ is true for all algebraically closed fields.
(f) Give an example of a field K and a map f showing that $\Psi(K, 1, 3)$ fails.
(g) Give an example of a surjective polynomial map $f : \mathbb{C} \to \mathbb{C}$ which is not injective.

31
Definable Sets and Varieties

In this chapter we fix an algebraically closed field K. The quantifier elimination theorem proved in Chapter 30 shows that every definable subset of K^n is quantifier-free definable. In this chapter we explore what the quantifier-free definable sets are, relating them to affine algebraic varieties and, more generally, to constructible sets.

31.1 Varieties

Quantifier-free formulas are, by definition, Boolean combinations of atomic formulas. The language L_{ring} has no relation symbols, so the only atomic formulas are of the form $t_1(\bar{x}) = t_2(\bar{x})$, where t_1 and t_2 are terms. L_{ring}-terms are interpreted in K as polynomials $f(\bar{x}) \in \mathbb{Z}[\bar{x}]$. If we allow *parameters* from a subfield F of K, that is, constant symbols naming the elements of F, then $L_{\text{ring}}(F)$-terms are interpreted as polynomials in $F[\bar{x}]$. So atomic formulas are of the form $f_1(\bar{x}) = f_2(\bar{x})$, for polynomials f_1 and f_2. Setting $f(\bar{x}) = f_1(\bar{x}) - f_2(\bar{x})$, the atomic formulas are all equivalent to formulas of the form $f(\bar{x}) = 0$.

Before moving to arbitrary Boolean combinations of these equations, we consider the positive Boolean combinations, that is, those built using conjunctions and disjunctions but not negations.

By the conjunctive normal form theorem (see Exercise 20.12), a positive Boolean combination of equations is equivalent to a formula of the form

$$\bigwedge_{i=1}^{r} \bigvee_{j=1}^{s_i} f_{ij}(\bar{x}) = 0.$$

Since K is a field, it has no zero divisors, so $\bigvee_{j=1}^{s_i} f_{ij}(\bar{x}) = 0$ if and only if $\prod_{j=1}^{s_i} f_{ij}(\bar{x}) = 0$. So writing $f_i(\bar{x})$ for the product $\prod_{j=1}^{s_i} f_{ij}(\bar{x})$, every positive quantifier-free formula is equivalent to a finite conjunction of polynomial equations $\bigwedge_{i=1}^{r} f_i(\bar{x}) = 0$.

The subsets of K^n defined by these systems of polynomial equations are known as varieties.

Definition 31.1 Let P be a set of polynomials from $K[x_1, \dots, x_n]$. The zero-set of P is $V(P) = \{\bar{a} \in K^n \mid \text{for all } f(\bar{x}) \in P,\ f(\bar{a}) = 0\}$. The subset $V(P) \subseteq K^n$ is called an *affine algebraic variety*, which we will abbreviate to *variety*. It is also called a *Zariski-closed subset* of K^n.

Remark 31.2 In algebraic geometry, the word *variety* is sometimes reserved for a Zariski-closed subset which is irreducible. See Exercise 31.3. Apart from affine varieties, there are also projective varieties, quasi-projective varieties, and abstract varieties. They can also be treated as definable sets, but we will keep to affine varieties.

In the definition of $V(P)$, the set P of polynomials does not need to be finite. So it appears that varieties are more general than positive quantifier-free definable sets. We will show that in fact they are the same thing.

Varieties are closely related to ideals of the polynomial ring. Given a subset $P \subseteq K[\bar{x}]$, we have defined the variety $V(P) \subseteq K^n$. We can also define an operation in the opposite direction.

Definition 31.3 Given any subset $S \subseteq K^n$, let $I(S)$ be the set of polynomials in $K[x_1, \dots, x_n]$ which vanish at all points in S. That is,

$$I(S) = \{f(\bar{x}) \in K[\bar{x}] \mid \text{for all } \bar{a} \in S,\ f(\bar{a}) = 0\}.$$

Lemma 31.4 *For all $S, S_1, S_2 \subseteq K^n$ and all $P, P_1, P_2 \subseteq K[x_1, \dots, x_n]$, the following hold:*

(i) *$I(S)$ is an ideal in $K[\bar{x}]$.*
(ii) *If $S_1 \subseteq S_2 \subseteq K^n$, then $I(S_1) \supseteq I(S_2)$.*
(iii) *If $P_1 \subseteq P_2 \subseteq K[x_1, \dots, x_n]$, then $V(P_1) \supseteq V(P_2)$.*
(iv) *$V(I(S)) \supseteq S$.*
(v) *$I(V(P)) \supseteq P$.*
(vi) *$I(V(I(S))) = I(S)$.*
(vii) *$V(I(V(P))) = V(P)$.*
(viii) *If I is the ideal generated by P, then $V(I) = V(P)$.*

The proof is left as an exercise. The characterisation of exactly which ideals are $I(S)$ for some S is given by Hilbert's Nullstellensatz, which is the subject of Chapter 32. With this relation between varieties and ideals, we can apply Hilbert's basis theorem.

Proposition 31.5 *Every variety $V \subseteq K^n$ is the zero set of a finite set of polynomials.*

Proof Suppose $V = V(P)$. Then, by Lemma 31.4, $V = V(I(V(P)))$. By Theorem 28.15, there is a finite set $\{f_1, \ldots, f_r\}$ of polynomials which generates $I(V(P))$. Using the lemma again, $V = V(f_1, \ldots, f_r)$. □

The following corollary summarises what we have proved.

Corollary 31.6 *Let K be an algebraically closed field and $n \in \mathbb{N}^+$. The varieties $V \subseteq K^n$ are exactly the subsets of K^n which are defined by positive quantifier-free formulas (using parameters from K).* □

31.2 Constructible Sets

Constructible sets are what geometers call subsets of K^n which are finite Boolean combinations of varieties. So they are exactly the subsets of K^n which are quantifier-free definable, with parameters from K. By quantifier elimination for algebraically closed fields, they are also exactly the definable sets. We will give a normal form for them and then explain how they can be considered as varieties.

By the disjunctive normal form theorem (see Exercise 20.12), every quantifier-free formula is equivalent to a formula of the form

$$\bigvee_{i=1}^{r} \left(\bigwedge_{j=1}^{s_i} f_{ij}(\bar{x}) = 0 \wedge \bigwedge_{j=1}^{t_i} g_{ij}(\bar{x}) \neq 0 \right)$$

for some natural numbers r, s_i, t_i and some polynomials $f_{ij}(\bar{x})$ and $g_{ij}(\bar{x})$. For each i, let $g_i(\bar{x})$ be the product $\prod_{j=1}^{t_i} g_{ij}(\bar{x})$, let V_i be the variety defined by $\bigwedge_{j=1}^{s_i} f_{ij}(\bar{x}) = 0$, and let W_i be the variety defined by $\bigwedge_{j=1}^{s_i} f_{ij}(\bar{x}) = 0 \wedge g_i(\bar{x}) = 0$. Then the set defined by the formula above is $\bigcup_{i=1}^{r} (V_i \smallsetminus W_i)$. Thus we have proved the following.

Proposition 31.7 *Let K be an algebraically closed field and $n \in \mathbb{N}^+$. Every subset of K^n which is L_{ring}-definable with parameters from K is a finite union of sets of the form $V \smallsetminus W$, where V and W are varieties in K^n.* □

It is generally much easier to deal with systems of equations than to deal with negations of equations as well. The *Rabinowitsch trick*, introduced in a one-page paper in 1930 [Rab30], turns the negated equations into equations, at the cost of introducing new variables. The observation is simply that the negated equation $g(x_1, \ldots, x_n) \neq 0$ is equivalent to the equation $x_{n+1}g(x_1, \ldots, x_n) = 1$, because in a solution $a_1, \ldots, a_{n+1}$ of this equation, $g(a_1, \ldots, a_n) = 1/a_{n+1}$, which cannot be 0.

Theorem 31.8 *Let $S \subseteq K^n$ be a constructible set. Then there is $r \in \mathbb{N}$ and a variety $Y \subseteq K^{n+r}$ such that the projection $p : K^{n+r} \to K^n$ onto the first n coordinates restricts to a bijection from Y to S.*

Proof As described above, we may assume that S is defined by a formula of the form

$$\bigvee_{i=1}^{r}\left(\bigwedge_{j=1}^{s_i} f_{ij}(x_1, \ldots, x_n) = 0 \wedge g_i(x_1, \ldots, x_n) \neq 0\right).$$

Take Y to be defined by

$$\bigvee_{i=1}^{r}\left(\bigwedge_{j=1}^{s_i} f_{ij}(x_1, \ldots, x_n) = 0 \wedge x_{n+i}g_i(x_1, \ldots, x_n) - 1 = 0\right).$$

This is a positive quantifier-free formula, so by Corollary 31.6, Y is a variety. It is immediate that p restricts to a bijection $Y \to S$. □

If we do not want the extra free variables, we can note that the projection is the same thing as existential quantification and get the following corollary.

Corollary 31.9 *Every definable subset of K^n is defined by a formula of the form $\exists x_{n+1} \ldots, x_{n+s} \left[\bigwedge_{i=1}^{s} f_i(x_1, \ldots, x_{n+s}) = 0\right]$, for some $s \in \mathbb{N}$ and some polynomials f_i.* □

31.3 Chevalley's Theorem

As an application of quantifier elimination for algebraically closed fields, we prove a theorem of Chevalley.

Theorem 31.10 (Chevalley's theorem) *Let K be an algebraically closed field, let S be a constructible subset of K^n, and let $f : K^n \to K^m$ be a polynomial map. Then the image $f(S) \subseteq K^m$ is a constructible set.*

Proof The polynomial map f has components $f_1, \ldots, f_m$, with each $f_i \in K[x_1, \ldots, x_n]$. The constructible set S is defined by a quantifier-free $L_{\text{ring}}(K)$

formula $\varphi(x_1, \ldots, x_n)$. The image $f(S)$ is defined in the free variables $y_1, \ldots, y_m$ by the formula

$$\exists x_1, \ldots, x_n \left[\varphi(x_1, \ldots, x_n) \wedge \bigwedge_{i=1}^{m} y_i = f_i(x_1, \ldots, x_n) \right].$$

By quantifier elimination for algebraically closed fields, Corollary 30.7, there is a quantifier-free $L_{\text{ring}}(K)$-formula $\theta(y_1, \ldots, y_m)$ (with the same parameters) which also defines $f(S)$. So $f(S)$ is constructible. □

Exercises

31.1 Suppose that $I_1 \subseteq I_2 \subseteq I_3 \subseteq \cdots \subseteq I_r \subseteq \cdots$ is an ascending chain of ideals of $F[\bar{x}]$. Using the Hilbert basis theorem, show there is $s \in \mathbb{N}^+$ such that for all $t > s$, $I_t = I_s$. [We say that $F[\bar{x}]$ has the *ascending chain condition* for ideals.]

31.2 Show that the collection of varieties in K^n satisfies the *descending chain condition*, that is, if

$$V_1 \supseteq V_2 \supseteq V_3 \supseteq \cdots \supseteq V_r \supseteq \cdots$$

is a descending chain of varieties then there is $s \in \mathbb{N}^+$ such that for all $t > s$, $V_t = V_s$.

31.3 A variety $V \subseteq K^n$ is said to be *irreducible* if it cannot be written as $V = V_1 \cup V_2$ for varieties V_1 and V_2, both proper subsets of V. Show that any variety is a finite union of irreducible varieties.

31.4 Show that a variety $V \subseteq K^n$ is irreducible if and only if $I(V)$ is a prime ideal of $K[\bar{x}]$.

31.5 Let $K \models \text{ACF}_p$ with $p > 0$. Show that the map $K \to K$ given by $x \mapsto \sqrt[p]{x}$ is a definable (single-valued) function. Deduce that if $F \subseteq K$ is a subfield, then a variety V is definable with parameters from F if and only if it is defined by polynomials with coefficients in the *perfect closure* of F, that is, in $F^{p^{-\infty}} = \left\{ a^{1/p^r} \mid a \in F, r \in \mathbb{N} \right\}$.

31.6 Let K be an algebraically closed field and $F \subseteq K$ be a subfield. Show that $b \in \text{acl}^K(F)$ if and only if b lies in a finite subset of K which is definable with parameters from F.

31.7 The property of the relative algebraic closure from Exercise 31.6 makes sense in any structure and is called the *(model-theoretic) algebraic closure*. Let $\mathcal{M}$ be any L-structure, and let $A \subseteq M$ be a subset and

$b \in M$. We define $b \in \mathrm{acl}(A)$ iff b is contained in a finite subset of $\mathcal{M}$ which is definable with parameters from A. Show that acl is a closure operator on M with finite character (see Exercise 30.3).

31.8 Let K be an algebraically closed field, and suppose that $S \subseteq K$ is definable (possibly with parameters). Show that S is either finite or cofinite, that is, $K \setminus S$ is finite.

31.9 A complete theory T such that the property of the previous exercise holds for every model of T is called *strongly minimal*. Show that if T is strongly minimal and $\mathcal{M} \models T$, then the closure operator acl on $\mathcal{M}$ is a pregeometry (see Exercise 30.3).

31.10 We outline Ax's generalisation of the result from Exercise 30.7. This is also known as the Ax–Grothendieck theorem.

(a) Write down an L_{ring}-sentence $\varphi_{n,t,d}$ which states that 'for any variety $V = V(g_1, \ldots, g_t) \subseteq K^n$ and any polynomial map $f : K^n \to K^n$, where the f_i and g_j have degree at most d, if f restricts to a map $V \to V$ and this map is injective, then it is surjective.'
(b) Use the method of Exercise 30.7 to prove that if K is any algebraically closed field and $n, t, d \in \mathbb{N}$, then $K \models \varphi_{n,t,d}$.
(c) Formulate and prove a similar statement for a constructible set S and a constructible map $f : S \to S$.

31.11 In this exercise we relate the Stone space of types over parameters to the Zariski spectrum of a polynomial ring.

Let K be an algebraically closed field, F a subfield, and $\bar{a} \in K^n$. Define $I_F(\bar{a}) = \{f(\bar{x}) \in F[\bar{x}] \mid f(\bar{a}) = 0\}$.

(a) Show that $I_F(\bar{a})$ is a prime ideal and that $\mathrm{tp}(\bar{a}/F) = \mathrm{tp}(\bar{b}/F)$ if and only if $I_F(\bar{a}) = I_F(\bar{b})$.
(b) Use this idea to construct a bijection from the Stone space $S_n(\mathrm{ACF}, F)$ of complete n-types with parameters from F to the Zariski spectrum $\mathrm{Spec}(F[x_1, \ldots, x_n])$, the set of all prime ideals of $F[x_1, \ldots, x_n]$.
(c) Show that $V(I_F(\bar{a}))$ is the smallest variety containing $\bar{a}$ which is defined by polynomials with coefficients from F.
(d) If you know about the topologies on the Stone space and the Zariski spectrum, prove that this map is continuous but that its inverse is not.

32
Hilbert's Nullstellensatz

In this final chapter we use quantifier elimination to prove Hilbert's Nullstellensatz (theorem on zero sets), which shows that the ideals corresponding to algebraic varieties are exactly the radical ideals. We prove the theorem in three stages to highlight the different techniques used in the proof. First, quantifier elimination is used to prove a theorem about systems of equations. Then we show that any ideal can be extended to a maximal ideal and thereby use the first theorem to prove the weak nullstellensatz. The strong Nullstellensatz can be proved in a similar way, but because the Rabinowitsch trick is so useful, we illustrate its use by deducing the strong Nullstellensatz from the weak version. This was actually Rabinowitsch's original application of the trick.

32.1 Systems of Polynomial Equations

The definition of algebraically closed fields only requires that each individual non-constant polynomial in one variable have a zero in the field. The power of the quantifier elimination theorem is that we can deduce a consequence for systems of polynomial equations (and negations of equations) in several variables.

Theorem 32.1 *Let K be an algebraically closed field. Then any finite system of polynomial equations and negations of equations with coefficients from K, in any number of variables, which has a solution in some field extending K already has a solution in K.*

Proof Let $\bar{c}$ be a tuple from K consisting of all the coefficients of all the polynomials in the finite system of equations and inequations. Then we can capture the system with an L_{ring}-formula $\varphi(\bar{x},\bar{c})$ where $\varphi(\bar{x},\bar{y})$ has the form

$$\bigwedge_{i=1}^{r} f_i(\bar{x},\bar{y}) = 0 \wedge \bigwedge_{i=1}^{s} g_i(\bar{x},\bar{y}) \neq 0$$

and the f_i and g_i are polynomials in $\mathbb{Z}[\bar{x},\bar{y}]$.

Suppose that B is a field extension of K such that $B \models \exists\bar{x}\varphi(\bar{x},\bar{c})$. Then $B^{\text{alg}} \models \exists\bar{x}\varphi(\bar{x},\bar{c})$ by Proposition 5.6.

Since ACF has quantifier elimination, there is a quantifier-free formula $\theta(\bar{y})$ such that $\text{ACF} \vdash \forall\bar{y}[\exists\bar{x}\varphi(\bar{x},\bar{y}) \leftrightarrow \theta(\bar{y})]$.

So $B^{\text{alg}} \models \theta(\bar{c})$ and, since θ is a quantifier-free formula and K is a substructure of B^{alg}, by Lemma 5.5, $K \models \theta(\bar{c})$. Then, since $K \models \text{ACF}$, we have $K \models \exists\bar{x}\varphi(\bar{x},\bar{c})$. □

This theorem says that to consider systems of polynomial equations, we do not have to look beyond an algebraically closed K to see if solutions are possible. However, it does not say explicitly which systems of equations do have solutions. For that we need Hilbert's Nullstellensatz.

32.2 The Weak Nullstellensatz

Theorem 32.2 (The weak Nullstellensatz) *Let K be an algebraically closed field, and let $I \subseteq K[\bar{x}]$ be an ideal. Then I is a proper ideal if and only if $V(I) \neq \emptyset$.*

Proof First note that I is a proper ideal if and only if it does not contain the constant polynomial 1. Since 1 does not vanish anywhere, if $1 \in I$, then $V(I) = \emptyset$.

Now suppose that I is a proper ideal. First we extend I to a maximal proper ideal J in $K[\bar{x}]$. To do this, index all the polynomials in $K[\bar{x}]$ as $(f_\alpha)_{\alpha<\lambda}$ for some ordinal λ. If K is countable, we can just index by natural numbers. Then set $I_0 = I$, and for each α in turn, set $I_{\alpha+1}$ to be the ideal generated by I_α and f_α if this ideal is proper, and set $I_{\alpha+1} = I_\alpha$ otherwise. For a limit ordinal β, take $I_\beta = \bigcup_{\alpha<\beta} I_\alpha$. Then we can take J to be I_λ.

Let $B = K[\bar{x}]/J$, which is a field extension of K because J is a maximal ideal. Let $\bar{b}$ be the image in B of $\bar{x}$ under the quotient map. Then, for each $f(\bar{x}) \in J$, so in particular for each $f(\bar{x}) \in I$, we have $f(\bar{b}) = 0$.

Using the Hilbert basis theorem, Theorem 28.15, there is a finite set $f_1,\ldots,f_r$ of polynomials which generates I. So $B \models \exists\bar{x} \bigwedge_{i=1}^{r} f_i(\bar{x}) = 0$.

By Theorem 32.1, $K \models \exists \bar{x} \bigwedge_{i=1}^{r} f_i(\bar{x}) = 0$. Let $\bar{a} \in K^n$ be a witness, so $\bar{a} \in V(I)$. So $V(I) \neq \emptyset$, as required. □

32.3 The Strong Nullstellensatz

Example 32.3 Let $f(x_1, x_2) = (x_1 - 6)$ and let $g(x_1, x_2) = f(x_1, x_2)^2$. Then $V(f) = V(g) = \{(6, x_2) \mid x_2 \in K\}$, but the ideals generated by f and g are different.

Definition 32.4 An ideal I of a ring R is called a *radical ideal* if, for all $m \in \mathbb{N}^+$, for all $f \in R$, if $f^m \in I$, then $f \in I$. The *radical* of an ideal I of a ring R is the ideal $\sqrt{I} = \{f \in R \mid \text{for some } m \in \mathbb{N}^+, f^m \in I\}$.

Lemma 32.5 *For any ideal $I \subseteq K[\bar{x}]$ we have $V\left(\sqrt{I}\right) = V(I)$.*

Proof Suppose $\bar{a} \in K^n$ and $f^m(\bar{a}) = 0$, that is, $f(\bar{a})^m = 0$. Then, since K is a domain, $f(\bar{a}) = 0$. □

So to understand varieties, we can restrict our attention to radical ideals. The next theorem, Hilbert's Nullstellensatz, shows that different radical ideals do correspond to different varieties. For this it is essential that the field K be algebraically closed. For example, in the real field $\mathbb{R}$, let $I_1 = \mathbb{R}[x]$ as an ideal of itself, and let $I_2 \subseteq \mathbb{R}[x]$ be the ideal generated by $x^2 + 1$. Then I_1 and I_2 are radical, but $V(I_1) = V(I_2) = \emptyset$. Taking the complex field $\mathbb{C}$ in place of $\mathbb{R}$, we have $V(I_1) = \emptyset$ and $V(I_2) = \{\pm i\}$.

Theorem 32.6 (Hilbert's Nullstellensatz) *Suppose that K is an algebraically closed field and $I \subseteq K[\bar{x}]$ is a radical ideal. Then $I(V(I)) = I$.*

Proof We will show that if I is any ideal of $K[x_1, \ldots, x_n]$ and $f \in I(V(I))$, then there is $m \in \mathbb{N}^+$ such that $f^m \in I$. So, in particular, if $I = \sqrt{I}$, then $f \in I$.

Using Hilbert's basis theorem, Theorem 28.15, choose a generating set $f_1, \ldots, f_r$ for I. Then $f \in I(V(I))$ means that for all $\bar{a} \in K^n$, if $f_i(\bar{a}) = 0$ for each $i = 1, \ldots, r$, then $f(\bar{a}) = 0$. Using the Rabinowitsch trick, we consider f and the f_i as polynomials in $x_1, \ldots, x_{n+1}$. Then $f_1, \ldots, f_r$ and $(1 - x_{n+1} f)$ have no common zeros in K^{n+1}. So, by the weak Nullstellensatz, Theorem 32.2, the ideal of $K[x_1, \ldots, x_{n+1}]$ generated by $f_1, \ldots, f_r$ and $(1 - x_{n+1} f)$ is not proper. So by Lemma 28.5, there are $g_0, g_1, \ldots, g_r \in K[x_1, \ldots, x_{n+1}]$ such that the equation

$$1 = g_0(1 - x_{n+1} f(x_1, \ldots, x_n)) + \sum_{i=1}^{r} g_i(x_1, \ldots, x_{n+1}) f_i(x_1, \ldots, x_n)$$

holds in the polynomial ring $K[x_1, \ldots, x_{n+1}]$.

We cancel the first term by substituting $1/f(x_1, \ldots, x_n)$ for the new variable x_{n+1} to get an equation

$$1 = \sum_{i=1}^{r} g_i(x_1, \ldots, x_n, 1/f(x_1, \ldots, x_n)) f_i(x_1, \ldots, x_n) \tag{32.1}$$

in the field $K(x_1, \ldots, x_n)$ of rational functions. Let m be the maximum degree in the variable x_{n+1} of the polyomials $g_1, \ldots, g_r$. Then we can write each $g_i(x_1, \ldots, x_n, 1/f(x_1, \ldots, x_n))$ as $\frac{h_i(x_1,\ldots,x_n)}{f(x_1,\ldots,x_n)^m}$ for some polynomials $h_i(x_1, \ldots, x_n) \in K[x_1, \ldots, x_n]$. Clearing denominators, Equation (32.1) becomes

$$f(x_1, \ldots, x_n)^m = \sum_{i=1}^{r} h_i(x_1, \ldots, x_n) f_i(x_1, \ldots, x_n))$$

in $K[x_1, \ldots, x_n]$, which by Lemma 28.5 says that $f^m \in I$, as required. □

The model theory of fields is now a large subject, and we have just scratched the surface. Introductions to various aspects of it can be found in [MMP06], [HV02], [Bou98], and [HHM08], as well as in the books mentioned earlier.

Exercises

32.1 Let K be an algebraically closed field, and let $I \subseteq K[\bar{x}]$ be an ideal. Show that $I = I(V(I))$ if and only if I is a radical ideal. Give an example to show that this fails if K is not algebraically closed.

32.2 Show that an ideal $I \subseteq K^{\mathrm{alg}}[\bar{x}]$ is maximal if and only if it is generated by polynomials of the form $x_1 - a_1, \ldots, x_n - a_n$. Give an example to show that this fails if K is not algebraically closed.

32.3 Give a complete proof of Hilbert's Nullstellensatz along the lines of the proof of Theorem 32.2, not using the Rabinowitsch trick.

32.4 Give a sketch proof of Hilbert's Nullstellensatz. Explain what all the key ideas in the proof are, how they fit together, and how they depend on ideas from earlier in the book.

Bibliography

[Ax69] James Ax. Injective endomorphisms of varieties and schemes. *Pacific J. Math.*, 31:1–7, 1969.

[Bal88] John T. Baldwin. *Fundamentals of Stability Theory*. Perspectives in Mathematical Logic. Springer, Berlin, 1988.

[BBR62] Andrzej Białynicki-Birula and Maxwell Rosenlicht. Injective morphisms of real algebraic varieties. *Proc. Amer. Math. Soc.*, 13:200–203, 1962.

[BCR98] Jacek Bochnak, Michel Coste, and Marie-Françoise Roy. *Real Algebraic Geometry*, volume 36 of *Ergebnisse der Mathematik und ihrer Grenzgebiete (3)* [Results in Mathematics and Related Areas (3)]. Springer, Berlin, 1998. Translated from the 1987 French original, revised by the authors.

[Bou98] E. Bouscaren, editor. *Model Theory and Algebraic Geometry*, volume 1696 of *Lecture Notes in Mathematics*. Springer, Berlin, 1998.

[Bri77] Jane Bridge. *Beginning Model Theory*. Oxford Logic Guides. Clarendon Press, Oxford, 1977.

[CK90] C. C. Chang and H. J. Keisler. *Model Theory*, volume 73 of *Studies in Logic and the Foundations of Mathematics*. North-Holland, Amsterdam, third edition, 1990.

[CL00] René Cori and Daniel Lascar. *Mathematical Logic*. Oxford University Press, Oxford, 2000. A course with exercises. Part I, Propositional calculus, Boolean algebras, predicate calculus, translated from the 1993 French original by Donald H. Pelletier, with a foreword to the original French edition by Jean-Louis Krivine and a foreword to the English edition by Wilfrid Hodges.

[Hal74] Paul R. Halmos. *Naive Set Theory*. Undergraduate Texts in Mathematics. Springer, New York, 1974. Reprint of the 1960 edition.

[HHM08] Deirdre Haskell, Ehud Hrushovski, and Dugald Macpherson. *Stable Domination and Independence in Algebraically Closed Valued Fields*, volume 30 of *Lecture Notes in Logic*. Association for Symbolic Logic, Chicago, IL, 2008.

[Hod93] Wilfrid Hodges. *Model Theory*, volume 42 of *Encyclopedia of Mathematics and Its Applications*. Cambridge University Press, Cambridge, 1993.

[Hod97] Wilfrid Hodges. *A Shorter Model Theory*. Cambridge University Press, Cambridge, 1997.

[HV02] Bradd Hart and Matthew Valeriote, editors. *Lectures on Algebraic Model Theory*, volume 15 of *Fields Institute Monographs*. American Mathematical Society, Providence, RI, 2002.

[Jec03] Thomas Jech. *Set Theory*. Springer Monographs in Mathematics. Springer, Berlin, Third millennium edition, revised and expanded. 2003.

[Kun11] Kenneth Kunen. *Set Theory*, volume 34 of *Studies in Logic (London)*. College Publications, London, 2011.

[Lan02] Serge Lang. *Algebra*, volume 211 of *Graduate Texts in Mathematics*. Springer, New York, third edition, 2002.

[Mar02] David Marker. *Model Theory: An Introduction*, volume 217 of *Graduate Texts in Mathematics*. Springer, New York, 2002.

[MMP06] David Marker, Margit Messmer, and Anand Pillay. *Model Theory of Fields*, volume 5 of *Lecture Notes in Logic*. Association for Symbolic Logic, La Jolla, CA, second edition, 2006.

[Pil83] Anand Pillay. *An Introduction to Stability Theory*, volume 8 of *Oxford Logic Guides*. Clarendon Press, Oxford, 1983.

[Pil96] Anand Pillay. *Geometric Stability Theory*, volume 32 of *Oxford Logic Guides*. Clarendon Press, Oxford, 1996.

[Poi00] Bruno Poizat. *A Course in Model Theory*. Universitext. Springer, New York, 2000. An introduction to contemporary mathematical logic, translated from the French by Moses Klein and revised by the author.

[Rab30] J. L. Rabinowitsch. Zum Hilbertschen Nullstellensatz. *Math. Ann.*, 102(1):520, 1930.

[Rot00] Philipp Rothmaler. *Introduction to Model Theory*, volume 15 of *Algebra, Logic and Applications*. Gordon and Breach, Amsterdam, 2000. Prepared by Frank Reitmaier, translated and revised from the 1995 German original by the author.

[Sac10] Gerald E. Sacks. *Saturated Model Theory*. World Scientific, Hackensack, NJ, second edition, 2010.

[She90] Saharon Shelah. *Classification Theory and the Number of Nonisomorphic Models*, volume 92 of *Studies in Logic and the Foundations of Mathematics*. North-Holland, Amsterdam, second edition, 1990.

[TZ12] Katrin Tent and Martin Ziegler. *A Course in Model Theory*, volume 40 of *Lecture Notes in Logic*. Association for Symbolic Logic, La Jolla, CA, 2012.

[Vää11] Jouko Väänänen. *Models and Games*, volume 132 of *Cambridge Studies in Advanced Mathematics*. Cambridge University Press, Cambridge, 2011.

[vdD98] Lou van den Dries. *Tame Topology and o-Minimal Structures*, volume 248 of *London Mathematical Society Lecture Note Series*. Cambridge University Press, Cambridge, 1998.

Index